Processing and Analysis of Hyperspectral Data

Edited by Jie Chen,
Yingying Song and Hengchao Li

Published in London, United Kingdom

IntechOpen

Supporting open minds since 2005

Processing and Analysis of Hyperspectral Data
http://dx.doi.org/10.5772/intechopen.78179
Edited by Jie Chen, Yingying Song and Hengchao Li

Contributors
Mbongowo Mbuh, Karvhari Kale, Mahesh Solankar, Dhananjay Nalawade, Rajesh Gogineni, Ashvini Chaturvedi, Xian-Hua Han, Keqiang Yu, Yanru Zhao, Yong He, Xiaoli Li, Nancy Arana-Daniel, Carlos Villaseñor, Javier Gomez-Avila, Alma Y. Alanis, Carlos Lopez-Franco

First published in London, United Kingdom, 2020 by IntechOpen
IntechOpen is the global imprint of INTECHOPEN LIMITED, registered in England and Wales, registration number: 11086078, 7th floor, 10 Lower Thames Street, London, EC3R 6AF, United Kingdom
Printed in Croatia

British Library Cataloguing-in-Publication Data
A catalogue record for this book is available from the British Library

Additional hard and PDF copies can be obtained from orders@intechopen.com

Processing and Analysis of Hyperspectral Data
Edited by Jie Chen, Yingying Song and Hengchao Li
p. cm.
Print ISBN 978-1-78985-109-0
Online ISBN 978-1-78985-110-6
eBook (PDF) ISBN 978-1-83880-462-6

For EU product safety concerns:
IN TECH d.o.o., Prolaz Marije Krucifikse Kozulić 3, 51000 Rijeka, Croatia, info@intechopen.com or visit our website at intechopen.com.

Meet the editors

Dr. Jie Chen is currently a professor at the Center of Intelligent Acoustics and Immersive Communications, School of Marine Science and Technology, Northwestern Polytechnical University. He has conducted research at the University of Nice Sophia-Antipolis, France, and the University of Michigan, Ann Abor, USA, from 2009 to 2014. His research interests include distributed optimization, stochastic signal processing, machine learning with applications to acoustic signal processing, and image processing. Dr. Chen served as Technical Program Co-Chair of the International Workshop on Acoustic Echo and Noise Control (2016), Distinguished Lecturer of the Asia-Pacific Signal and Information Processing Association (2018–2019), and Program Co-Chair of the Intelligent Signal and Information Processing Summer School (2019) of the IEEE Signal Processing Society.

Dr. Yingying Song received her Dipl-Ing degree in System, Network and Telecommunication Engineering from the University of Technology of Troyes, France, in 2015, and her PhD degree from the Université de Lorraine, France, in 2018. She has worked at the Centre de Recherche en Automatique de Nancy, University of Lorraine, France. Her current research interests include hyperspectral image deconvolution, adaptive image processing, and hyperspectral image unmixing.

Dr. Hengchao Li is currently a professor at the Sichuan Provincial Key Laboratory of Information Coding and Transmission, Southwest Jiaotong University, Chengdu, China. He was a visiting scholar at the University of Colorado at Boulder, Boulder, CO, USA, during 2014. His research interests include statistical analysis of synthetic aperture radar images, remote sensing image processing, and signal processing in communications. Dr. Li was awarded the New Century Excellent Talents in University from the Ministry of Education of China in 2011. In addition, he has also been a reviewer for several international journals and conferences. He is currently serving as an associate editor of the *IEEE Journal of Selected Topics in Applied Earth Observations and Remote Sensing*.

Contents

Preface

Hyperspectral imagery has received considerable attention in the last decade as it provides rich spectral information and allows the analysis of objects that are unidentifiable by traditional imaging techniques. It has a wide range of applications, including remote sensing, industry sorting, food analysis, biomedical imaging, etc. However, in contrast to RGB images from which information can be intuitively extracted, hyperspectral data is only useful with proper processing and analysis. This emphasizes the importance of using advanced signal processing, image processing, and machine learning techniques for such a purpose. Classical hyperspectral image analysis tasks include target detection, classification, and spectral unmixing. This book intends to provide a comprehensive overview of the recent state of the art of these tasks. Thereafter, considering the prosperous study of deep-learning-based image and data analysis, this book also aims to collect the latest results of hyperspectral data analysis that benefit from deep neural networks. Finally, practical applications will be included to show how these analyses are useful in promoting real industry, medical, and biological development.

The book covers two sections, namely, Theoretical Advances of Hyperspectral Image Processing and Applications of Hyperspectral Image Processing. In the first section, the chapters "Hyperspectral Endmember Extraction Techniques" and "Hyperspectral Image Classification" present typical techniques, both classical and deep-learning based, for unmixing and classification tasks. The chapter "Hyperspectral Image Super-Resolution Using Optimization and DCNN-Based Methods" presents optimization-based and deep-learning-based super-resolution techniques. The chapter "Fast Chaotic Encryption for Hyperspectral Images" considers another fundamental but important aspect, i.e., the encryption of data. The second section includes two application-oriented chapters. Hyperspectral techniques are used for evaluating the quality of tea and water, respectively, in "NIR Hyperspectral Imaging for Mapping of Moisture Content Distribution in Tea Buds During Dehydration" and "Use of Hyperspectral Remote Sensing to Estimate Water Quality." The editors believe that readers can benefit from these chapters and gain a better understanding of hyperspectral techniques.

Jie Chen
Centre of Intelligent Acoustic and Immersive Communications,
School of Marine Science and Technology,
Northwestern Polytechnical University,
China

Yingying Song
Centre de Recherche en Automatique de Nancy (CRAN),
CNRS, University of Lorraine,
France

Hengchao Li
Sichuan Provincial Key Laboratory of Information Coding and Transmission,
Southwest Jiaotong University,
China

Section 1

Theoretical Advances of Hyperspectral Image Processing

Chapter 1

Hyperspectral Endmember Extraction Techniques

Karbhari V. Kale, Mahesh M. Solankar and Dhananjay B. Nalawade

Abstract

Hyperspectral data processing and analysis mainly plays a vital role in detection, identification, discrimination and estimation of earth surface materials. It involves atmospheric correction, dimensionality reduction, endmember extraction, spectral unmixing and classification phases. One of the ultimate aims of hyperspectral data processing and analysis is to achieve high classification accuracy. The classification accuracy of hyperspectral data most probably depends upon image-derived endmembers. Ideally, an endmember is defined as a spectrally unique, idealized and pure signature of a surface material. Extraction of consistent and desired endmember is one of the important criteria to achieve the high accuracy of hyperspectral data classification and spectral unmixing. Several methods, strategies and algorithms are proposed by various researchers to extract the endmembers from hyperspectral imagery. Most of these techniques and algorithms are significantly dependent on user-defined input parameters, and this issue is subjective because there is no standard specificity about these input parameters. This leads to inconsistencies in overall endmember extraction. To resolve the aforementioned problems, systematic, generic, robust and automated mechanism of endmember extraction is required. This chapter gives and highlights the generic approach of endmember extraction with popular algorithm limitations and challenges.

Keywords: hyperspectral imaging, endmember extraction, spectral signatures, spatial features, spectral unmixing, supervised classification

1. Introduction

The hyperspectral imaging (HSI) is a renowned technology that uses the chemical composition based spectroscopic properties of earth surface materials for their detailed analysis and exploration [1]. HSI is spectrally overdetermined and has the ability to observe the surface materials continuously across the wide range of electromagnetic spectrum, generally covering 0.4–2.5 μm, where different regions of wavelengths gives information about different material contents [2]. HSI, rather than comprehensively depending only on the spatial variations within the image (as in multispectral imaging), has moved to take the advantage of spectral variations for detailed material analysis [3]. The narrow and continuous spectral measurement provides the enriched amount of information significantly usable for target detection, material identification, material mapping, surface material classification,

IntechOpen

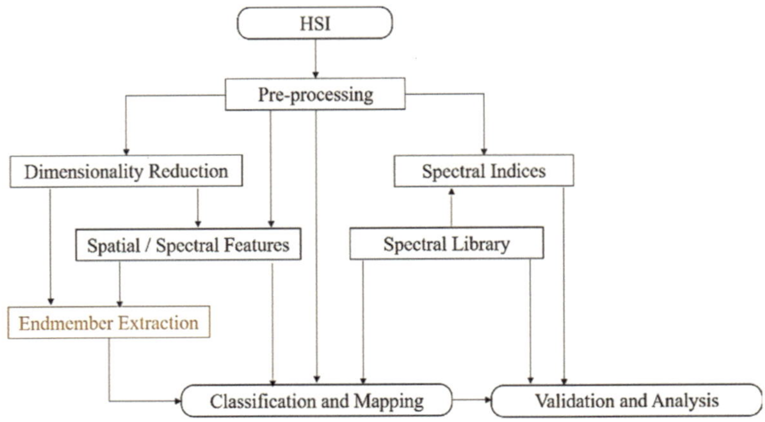

Figure 1.
Generalized hyperspectral data processing framework.

abundance estimation and mapping details of surface properties [2, 4]. These capabilities of HSI led to extend its areas of application and actively appealing researchers coming from varying subject domains to solve their respective problems. The popular areas, where hyperspectral imaging is providing its significance includes agriculture, forestry, hydrology, oceanography, soil analysis, land use land cover mapping and geology, etc.

Along with the ample amount of information, the hyperspectral imaging is characterized by several data processing and analysis challenges including adverse atmospheric effects, curse of dimensionality, unavailability of the ground-truth information and spectral mixing [1–3]. To extract the maximum information from this high volume HSI data with greater precision, these challenges needs to be handled efficiently. Of course, there several popular digital image processing and pattern recognition techniques are well available in the literature, but those techniques cannot be used as it is to process HSI data due to its different challenges. To process and analyses the hyperspectral data proficiently with better accuracy, there is a need to follow the generalized and application independent data processing framework (**Figure 1**) [1].

The ultimate intention of hyperspectral image analysis is to achieve the better classification or mapping accuracy, but it is adversely affected due the two important issues. First, the unavailability of the ground truth information or lower number of training samples leads to the Hughes effect, which decreases the classification accuracy as the number of features increases. Second, the spectral mixing problem raised either due to the limited spatial resolution of the sensor or homogenously combined mixture of surface materials. This leads to the linearly combined spectral reflectance of the component materials for every mixed pixel and keep that particular pixel away from any of the class in classification. To resolve these significant obstacles of HSI data classification and mapping, the term, image derived "endmember" comes into picture and contribute with remarkable role in increasing HSI data classification accuracy [1].

2. Hyperspectral endmembers

Due to significantly enhanced spatial and spectral resolutions of the hyperspectral imaging sensors, hyperspectral endmember extraction (EE) is found to be a significant step in hyperspectral data analysis. As described in [5], "a hyperspectral endmember (also known as 'pure pixel') is an idealized, pure signature of a spectral class." The pure spectral signature signifies the complete reflectance of pixel exclusively occupied by a single surface material. In case of multispectral imagery, finding endmembers within the scene is significantly difficult because of the broad spectral bands with lower spatial and spectral resolutions. Therefore the EE is not taken as an important step in multispectral data analysis and its importance has remained unnoticed. Thereafter, with the substantial improvements in hyperspectral sensors several subtle surface material contents that cannot be explored by multispectral imaging can now be explored using the hyperspectral imagery. These material contents are usually not known previously and can be only analyzed with the higher spectral resolution data. The hyperspectral endmembers are supposed to be one of these material contents. In HSI data, finding endmembers is a critical task due to their population within the data is significantly low and generally beyond the human visual perception [6]. Once the endmembers are extracted precisely from the HSI image itself, they can be used as a reference data or training data to enhance the classification accuracy or efficient spectral unmixing of the hyperspectral images [1]. These image derived endmembers are more effective as compared to the standard laboratory or field spectra, as they are not certainly recorded under the similar conditions as the satellite or airborne HSI data.

There are several EE techniques and methods are available in the literature, each having its own advantages and challenges. Depending on the EE approaches, these techniques are broadly classified into two categories (**Figure 2**). First, the convex geometry based EE. Further, it is sub-classified into the Orthogonal Projection (OP) based and Simplex Volume (SV) based EE approaches. The OP based EE approaches, makes the orthogonal projection of all data samples onto a set of selected vectors and considers the data samples producing extreme (either minimal or maximal) projections with these selected vectors as a final set of endmembers. The popular OP based EE algorithms are Pixel Purity Index (PPI) [7], VCA [8] and Sequential Maximum Angle Convex Cone (SMACC) [9]. The SV based EE approaches, assumes that the simplex formed by a set of pure signatures as vertices should produce the maximum volume among all simplexes formed by the same number of signatures as vertices. The SV based EE algorithms are NFINDR [10], Simplex Growing Algorithm (SGA) [11] and Convex Cone Analysis (CCA) [12]. Second, the statistics based EE. Further, it is sub-classified into second order statistics based EE, higher order statistics based EE and Swarm Intelligence based EE. The Iterative Error Analysis (IEA) [13] and EESCA [14] techniques comes under

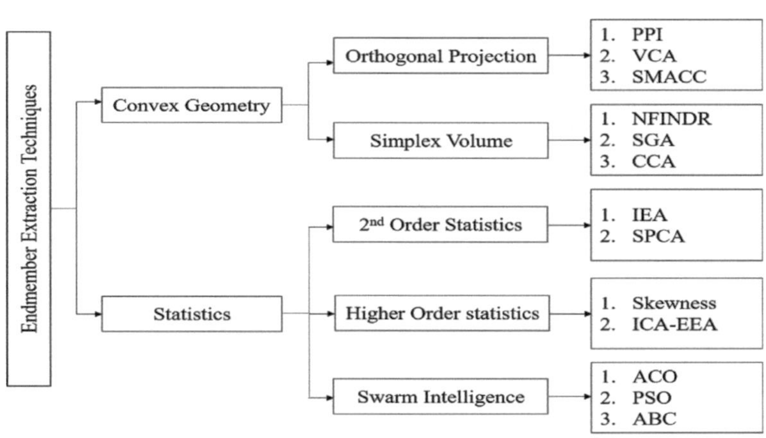

Figure 2.
EE techniques.

the category of second order statistics based EE. The skewness and Independent Component Analysis EE algorithm (ICA-EEA) [6] comes under the category of higher order statistics based EE. Apart from these traditional techniques, few significant Swarm Intelligence (SI) based approaches are also discussed. The Ant Colony Optimization (ACO) [15], Particle Swarm Optimization (PSO) [16] and Artificial Bee Colony (ABC) [17] and others are the typical examples of SI based EE algorithms. SI is the new area in artificial intelligence and deals with the natural and artificial systems composed of multiple entities that are articulated using self-organizations and decentralized control. In particular, the discipline emphases on the mutual behaviors that resulted from the interactions of entities with each other & their environment. SI algorithms have capability to solve the problem of combinatorial optimization.

In the further portion, this chapter present and discuss these EE algorithms in step-wise manner along with few practical experiments. The Samson hyperspectral image (**Figure 3**) is used for practical demonstrations. This image contains 156 spectral channels (having 3.13 nm spectral resolution) covering from 401 to 889 nm range of electromagnetic spectrum. Spectrally, there are three classes are available into this image, i.e., soil, tree and water [18].

2.1 Pixel purity index (PPI)

The PPI algorithm [7] is a mostly used endmember extraction algorithm due to its availability into the ENVI commercial software package. Its step-by-step implementation of is never been revealed into the literature due to its propriety rights. Considering its basic idea, several endmember extraction algorithms [19–21] tried to mimic the implementation of original PPI [7] and produced the useful results. The PPI is an orthogonal projection based endmember extraction mechanism and works through stages briefly explained below:

2.1.1 Initialization

The algorithm initiates with the randomly generate "k" unit vectors, also called as "skewers," $\{skewer_k\}_{k=1}^{k}$ where k is a supposed to be user defined large positive integer.

2.1.2 PPI count calculation

In PPI count calculation, every data sample is orthogonally projected onto all the skewers $\{skewer_k\}_{k=1}^{k}$. Further, for every data sample **r**, identify those skewers onto

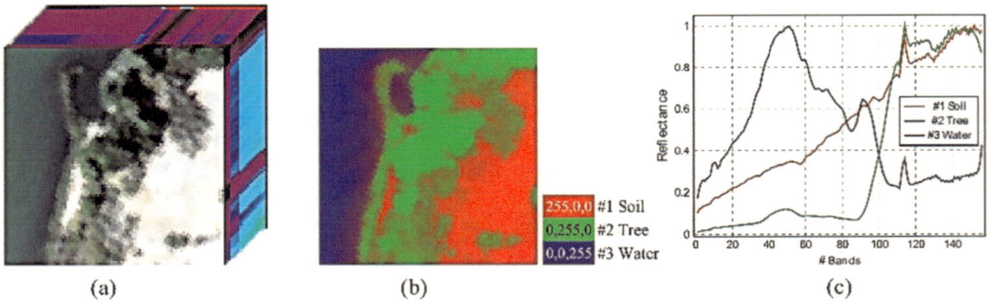

Figure 3.
Samson hyperspectral image with ground truth (GT). (a) Samson. (b) GT: abundances. (c) GT: endmembers.

which data sample **r** produces extreme (either the maximal or minimal) projection at their end points. These skewers producing extreme projections, form a set (i.e., $S_{extrema}(r)$) for that particular data sample **r**. Then, the PPI count of that particular data sample **r** is equal to the cardinality (i.e., total number of skewers within the set) of the set $S_{extrema}(r)$, that is,

$$n_{PPI}(r) = |S_{extrema}(r)| \tag{1}$$

where |A| is defined as the cardinality of set A.

2.1.3 Candidate selection

In candidate selection, the user defined value is set for threshold t to threshold the PPI count $n_{PPI}(r)$ for all data sample vectors to select the possible data samples for endmembers.

2.1.4 Endmember extraction

At the end, all the data samples having $n_{PPI}(r) \geq t$ as considered as final set of endmembers.

Figure 4 enlightens the endmembers extracted using PPI from Samson data. In this experiment, PPI is executed with 1000 iterations and 50 as a threshold value, where it has extracted seven signatures as a final set of endmembers. Out of these seven endmembers, one represents the "soil", four represents the "trees" and two represents the "water." One endmember signature (along with their respective spatial coordinates) for each class in the Samson image is given in **Figure 4** along with ground truth spectra's.

Although PPI is mostly used endmember extraction algorithm, but initially it was not considered as a solution for endmember extraction, but it was considered as a guide for endmember extraction. The PPI is very sensitive to the value of randomly generated skewers and threshold. The need of large number of randomly generated skewers leads to the higher computations complexity and yields varying outcomes (due to randomly generated skewers) during several runs even on the same hyperspectral image. Further the undefined criteria for selecting the appropriate value of threshold put forward the need of trained user for selection the final set of endmembers [19]. To deal with these challenges, there are few attempts available into the literature that tried to improve the performance of PPI with some modifications at algorithm level. In [19, 20], the attempt is made use the concept of Virtual Dimensionality (VD) to automatically identify the number of endmembers to be identified from the image and opted the algorithm based endmember initialization to produce the appropriate initial endmember set, which minimizes the

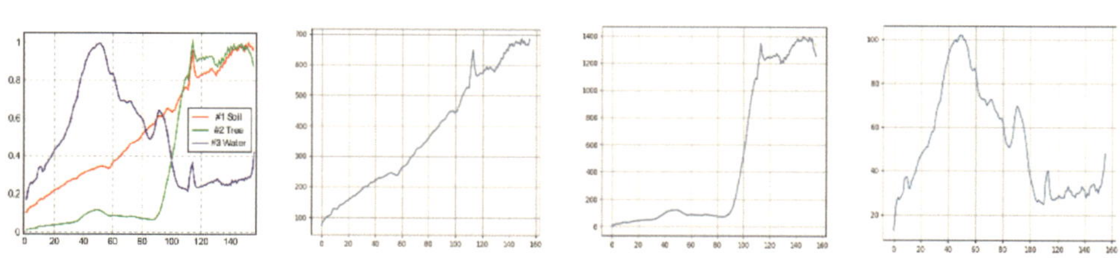

GT | Soil (69,29) | Tree (4,84) | Water (41,9)

Figure 4.
PPI extracted endmember from Samson data.

large number of runs required for PPI. In [21], multi-dimensional PPI (MDPPI) is presented as a fast alternative for PPI, which iteratively identifies the convex hull indices in lower-dimensional random projections.

2.2 Vertex component analysis (VCA)

The vertex component analysis (VCA) algorithm [8, 22, 23] is an unsupervised endmember extraction algorithm and works with the assumption that, in linear spectral mixing, every pixel signature is composed with the linear combinations of endmember spectra available within the scene. The VCA algorithm explores two facts: One, the endmembers are found to be the vertices of the simplexes and two, the affine transformation of every simplex is too simplex. The VCA algorithm initiates with the assumption that, there is a presence of endmembers within the data and iteratively projects data sample vectors on to the direction orthogonal to the subspace covered by the hyperspectral endmembers which are already identified. The new endmembers relates to the extreme (either minimal of maximal) projections. The algorithm continues to iterate till the number of endmembers are exhausted. The step-wise VCA algorithm is briefed below:

a. Suppose r be the $L \times 1$ vector, where L represents the total number of spectral channels, and m_i is the spectral signature of ith endmember, therefore

$$r = e\alpha, \tag{2}$$

where, $e = [e_1, e_2, \cdots, e_p]$ and $\alpha = [\alpha_1, \alpha_2, \cdots, \alpha_p]^T$ is the abundance portion of every endmember, and p is the total number of endmembers available within the image.

b. Due to the non-negativity constraint, the endmember abundance fractions satisfy the conditions given in Eqs. (3) and (4).

$$0 \leq \alpha_k \leq 1 \tag{3}$$

$$\sum_{k=1}^{p} \alpha_k = 1 \tag{4}$$

c. Every data sample can be considered as a sample vector within the L-dimensional Euclidean space, where every channel is allocated to the one axis of space and all are mutually orthogonal. Because of the constraints in Eqs. (3) and (4), the observed sample vector r belongs to the simplex having endmembers in the vertices. The detailed implementation and evaluation of VCA is given in [8].

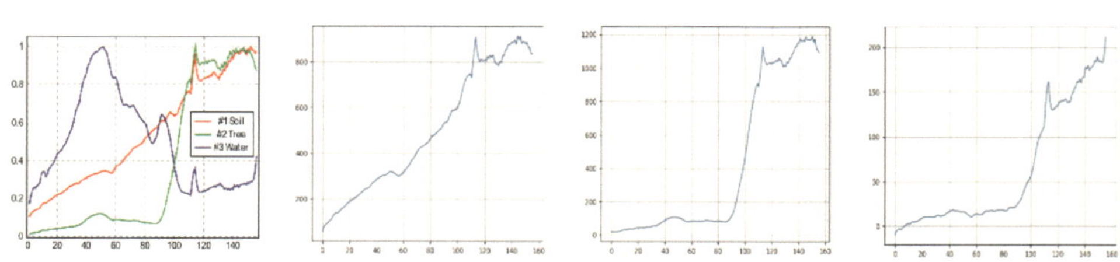

GT | Soil (69,28) | Tree (4, 84) | Unknown (17,54)

Figure 5.
VCA extracted endmember from Samson data.

Figure 5 gives the endmembers extracted using VCA from Samson data. In this experiment VCA is executed to extract three endmembers using nine iterations. It has extracted three endmembers with spatial coordinates (69, 28), (4, 84) and (14, 54), out of which, first signature represents the "soil," second signature represents the "tree" and third signature represents the unknown or may be mixed pixel.

The VCA algorithm is considered as a sequential implementation of the PPI algorithm. It is having lower computations cost as compared to the PPI and NFINDR EE algorithms. The usage of VCA is requires the precise knowledge about the number of endmembers to be identified. Its random initialization nature leads to the inconsistent outcomes during several executions even of the same image.

2.3 Sequential maximum angle convex cone (SMACC)

The sequential maximum angle convex cone (SMACC) [9] endmember extraction algorithm make use of a convex cone model along with non-negativity or sum-to-unity constraints to identify hyperspectral endmembers. In SMACC, the extreme projection points are used to determine a convex cone, which describes the first endmember. Further the constrained oblique projection is applied to the existing cone to extract another endmember. The convex cone is enlarged to comprise the new endmember. This procedure is revised till a projection extracts the endmember that is already available in the convex cone or till the defined number of endmembers is extracted.

In other words, SMACC algorithm first identifies the brightest pixel from the image; thereafter, it identifies the pixel mostly diverse from the brightest. Thereafter, it identifies the pixel that is most diverse as compared to the first two pixels. This procedure is revised until SMACC identifies a pixel which is already accounted for within the group of the formerly identified pixels, or until it identifies a previously defined number of endmembers.

As the convex approaches that are dependent on a simplex analysis, the number of hyperspectral endmembers is not limited by the total number of spectral bands. Although the endmembers identified by SMACC are exclusive, a one-to-one correspondence is not available among the total number of materials within the image and the total number of endmembers. SMACC extracts endmembers from pixels within the image. Every image pixel may be occupied by only single material or may be by high proportion of a single material with unique combinations of other surface materials. Every single material identified from the image is defined by a subset covering its spectral inconsistency. SMACC algorithm provides the endmember basis that describes each of these material subsets. Apart from this, SMACC also offers abundance images to identify the fractions of the total spectrally integrated pixels contributed by each constituent endmembers. Mathematically, SMACC algorithm uses the following convex cone expansion for every endmember spectra, defined as p:

$$p(c,i) =_{k}^{N} R(c,k)A(k,j) \tag{5}$$

where, i indicates the pixel index, j and k represents the endmember indices from 1 to the expansion length N, R is the matrix containing endmember signatures as columns, c gives the spectral channel index and A gives the matrix that comprises the proportional contribution of every individual endmember j in every endmember k for every pixel.

The SMACC EE tool available into the ENVI software used for EE from Samson data. The SMACC is executed to extract three endmembers. As compared to the GT

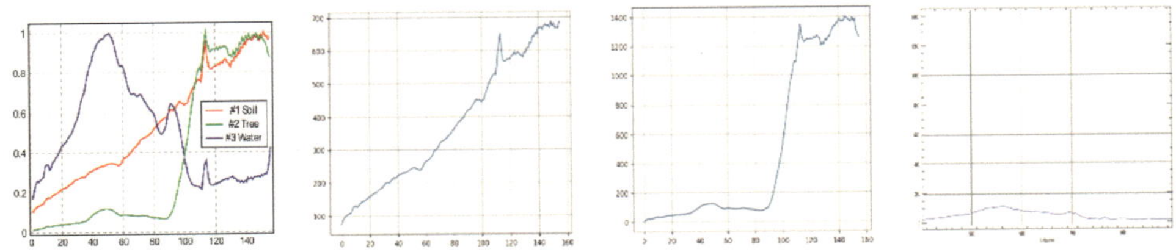

GT Soil Tree Unknown

Figure 6.
SMACC extracted endmembers from Samson data.

in **Figure 6**, SMACC has successfully extracted two endmembers signifying the "soil" and "tree" classes. The third signature belongs to the unknown class, which is incorrectly reported as an endmember by SMACC.

The overall observation is that, SMACC provides a faster and more automated method for finding spectral endmembers, but it is more approximate and yields less precision.

2.4 NFINDR

The NFINDR [10] algorithm is basically based on the geometric properties of the convex sets. The fundamental idea of NFINDR is to identify data samples that can produce a simplex having maximum volume and these samples are considered as final endmembers. Wherein, the algorithm assumes that, in L-spectral dimensions, the L-volume contained by a simplex produced of the purest data sample is bigger as compared to any other volume produced by any other combinations of data samples. The algorithm proceeds by "inflating" a simplex within the data, initializing with the randomly selected set of pixels. For every data sample and for every endmember, the endmember is substituted with the spectra of the data sample and the volume is re-calculated. If the volume founds increased, the spectra of the new data sample substitutes that endmember. This procedure is revised till no any replacement takes place. The step-wise implementation along with the mathematical perspective is briefly given below:

Usually, the spectrum of a given data sample is supposed to be linear combinations of the endmember signatures.

$$p_{ij} = \sum_k e_{ik} c_{kj} + \varepsilon \tag{6}$$

where p_{ij} indicates the ith spectral channel of jth data sample, e_{ik} indicates the ith spectral channel of kth pure pixel, c_{kj} denotes the proportions of spectral mixing for the jth data sample from the kth pure pixel and ε indicates the Gaussian error assumed to be very small. As the sample compositions in spectral mixing are expected to be in percentage, all the mixing proportions must sum to one:

$$\sum_k c_{kj} = 1 \tag{7}$$

If, $c_{kj} \approx 1$ for any of the endmember contribution within the data sample, the other pure pixel contribution is considered nearly equal to zero and the data sample can be considered as an endmember.

2.4.1 Preprocessing

In preprocessing, the orthogonal subspace projection (OSP) and maximum noise fraction (MNF) transformations are applied to the dataset to minimize the data dimensionality one less than the total number of endmembers to be identified. It helps to determine the data sample volume in a simpler manner.

2.4.2 Volume determination

Consider E as an endmember matrix augmented by row of ones:

$$E = \begin{bmatrix} 1 & 1 & \cdots & 1 \\ \vec{e}_1 & \vec{e}_2 & \cdots & \vec{e}_i \end{bmatrix} \tag{8}$$

where $\vec{e_i}$ is the column vector comprising the spectrum of endmember i.

Now, the simplex volume (V) produced with the endmember estimates is directly proportional to determinant E.

$$V(E) = \frac{1}{(l-1)!} abs(|E|) \tag{9}$$

Here, $(l-1)$ represents the number of dimensions occupied by data.

2.4.3 Endmember selection

The algorithm is stopped when all the pixels are tested and no replacement takes place.

Figure 7 gives the endmembers extracted using NFINDR from Samson data. In this experiment NFINDR is executed to extract three endmembers using nine iterations. It has successfully extracted three endmembers with spatial coordinates ((69, 29), (4, 84) and (1, 1)), one representing each class as per the GT.

Though this algorithm is mostly referred into the literature, there were no well-defined criteria for identifying the number of endmembers to extract. This challenge is tried to recover in [24] by using the notion of VD for defining the number of endmember available within the scene. Another issue with N-FINDR was its random initialization nature, which not only affects the algorithm convergence rate but also affects the final outcomes. There are several modifications are done in N-FINDR and made available in various versions [6, 25].

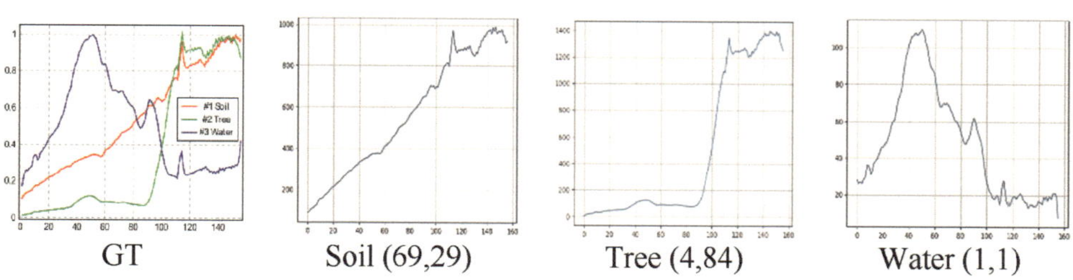

Figure 7.
NFINDR extracted endmembers from Samson data.

2.5 Simplex growing algorithm (SGA)

The simplex growing algorithm (SGA) [11] is a simplex based sequential pure pixel identification mechanism. It identifies the simplex having the maximum volume each time the new vertex is added. It is also called as the modified version of N-FINDR algorithm. The step-wise SGA is elaborated below:

2.5.1 Initialization

The SGA uses the VD estimation count to define the number of endmembers p to generate; and uses the e_1 identified by the very first pure pixel selection procedure as the preferred initial pure pixel and initialize n = 1.

At $n \geq 1$ and for every data sample r, we compute the volume $V(e_1, \cdots, e_n, r)$ defined by equation

$$V(e_1, \cdots, e_n, r) = \frac{\left| \det \begin{bmatrix} 1 & 1 & \cdots & 1 & 1 \\ e_1 & e_2 & \cdots & e_n & r \end{bmatrix} \right|}{n!} \tag{10}$$

It is the volume of simplex spanned by the vertices $e_1, e_2, \cdots, e_n, r$, symbolized with $S(e_1, e_2, \cdots, e_n, r)$. As the matrix $\begin{bmatrix} 1 & 1 & \cdots & 1 & 1 \\ e_1 & e_2 & \cdots & e_n & r \end{bmatrix}$ in Eq. (10) is not essentially a square shaped matrix, the efficient DR technique like principal component analysis (PCA) or maximum noise fraction (MNF) is used to minimize the data dimensionality L to dimension n.

Identify the e_{n+1} that produces the maximum of Eq. (10) and given by Eq. (11) given below:

$$e_{n+1} = \arg\left\{ \max_r \left[V(e_{e1}, \cdots, e_n, r) \right] \right\} \tag{11}$$

Stopping rule: if $n < p$, then $n \leftarrow n + 1$ and go to the step b. If not, the final set of vertices given by $\{e_1, e_2, \cdots, e_p\}$ is considered as the p number of desired endmembers.

This is the step-wise process of SGA to find endmembers from hyperspectral images. The SGA algorithm efficiently solves the three significant challenges observed in N-FINDR algorithm. First, SGA performs the automated estimation of number of pure pixels to find, second, it produces the consistent final set of endmembers and third, it performs efficiently with less computational complexity. With the coincidence, the basic idea of using growing simplexes to find endmembers in SGA is analogous to the that used in VCA, but their corresponding approached are dissimilar. Most specifically, the VCA algorithm is unable to resolve the second challenge and produces inconsistent outcomes due to its random initialization nature [11].

2.6 Convex cone analysis (CCA)

The convex cone analysis (CCA) [12] algorithm assumes that physical quantities like radiance of reflectance of the hyperspectral images are always positive. The sample vectors produced by the discrete radiance or reflectance spectra can be articulated as linear mixtures or combinations of non-negative components, which reside within the nonnegative also called as convex region. The fundamental objective of CCA algorithm is to identify the boundary points for that convex region. The theoretical implementation of CCA algorithm is discussed below:

a. To practically implement this idea, the algorithm computes the eigenvectors of the data sample correlation matrix of the scene, and only chooses those eigenvectors conforming to the *E* largest eigenvalues (wherein, *E* is a previously defined number of endmembers to extract).

b. The algorithm then looks forward to identify the boundary points of the convex cone, wherein the linear combinations of these eigenvectors forms sample vectors that are exactingly non-negative, by using the equation below:

$$h(x,y) = p_1 + a_1 p_2 + \cdots + a_{E-1} p_E \geq 0 \tag{12}$$

where $h(x,y)$ represents the spectral signature at pixel having spatial coordinates (x, y), the p_i indicates the eigenvectors belonging to the largest eigenvalues, and 0 indicates the zero vector.

c. These identified points characterize the corner points of the convex cone region. Then, Eq. (6) is rewritten as:

$$h(x,y) = [p_1 \cdots p_E] \begin{bmatrix} 1 \\ a_1 \\ \cdots \\ a_{E-1} \end{bmatrix} = P_a \geq 0 \tag{13}$$

wherein, the p_i are the N-dimensional column vectors. For *N*>*E*, $P_a = 0$ is an overdetermined system of the linear equations. If the elements of *P* are considered as the coefficients, and elements of ***a*** are considered variables, then, there are ***N*** equations of the given form:

$$p_{j1} + a_1 p_{j2} + \cdots + a_{E-1} p_{jE} = 0, \text{ for } j = 1, \cdots, N \tag{14}$$

That defines (*E*-1)-dimensional hyperplane within the *E*-dimensional space.

d. Precise endmember set can be found by using the (*E*-1)-tuples from the *N* equations. These solutions will form the linear combinations of eigenvectors that is having minimum *E*-1 zeros.

e. At the end, the boundary points of the convex cone region are considered as the endmember that satisfy Eq. (12) or equivalently, Min[h (x,y)] = 0, where the smallest is taken over all $h_i \in h(x,y), i = 1, \cdots, N$.

These sample vectors can be coined as final set of endmembers.

2.7 Endmember extraction using sparse component analysis (EESCA)

The endmember extraction using sparse component analysis (EESCA) [14] uses the sparse characteristics of abundance for endmember extraction from hyperspectral imagery. This algorithm is basically dependent of the primary endmember set of the VCA algorithm and thereafter two noteworthy iterative approaches are followed. The first one used to modify the endmember matrix and second one is to improve the pure pixel signature to enhance the algorithm accuracy. The step-wise mechanism of EESCA algorithm is summarized below:

The algorithm takes hyperspectral image as an input in the form of $X = [x(1), x(2), \ldots, x(T)] \in \mathbb{R}^{L \times T}$ and proceeds further.

a. Initiate the endmember matrix, i.e., $A^{[0]}$ using VCA algorithm.

b. Modify the pure pixel signatures using hyper-line estimation as per the steps given below:

1.Set the value of error tolerance, i.e., ε, and repeat $k = 1 : K$

2.Compute the distance $d\left(x(t), a_j^{[k-1]}\right)$ of all the data samples using Eq. (4) of [14], and allocate them in dissimilar classes$\left\{\Omega_j^{[k]}, j = 1, \cdots, p\right\}$.

3.Choose a subset matrix of M pixels from $\Omega_j^{[k]}$. Now, use SVD to modify the value of $a_i^{[k]}$ and eigenvalue$\lambda_i^{[k]}$.

4.If $\left|\lambda_j^k - \lambda_j^{[k-1]}\right| \le \varepsilon$, the pure pixel signature $a_j^{[k]}$ will be stable and not modified in the upcoming iteration.

c. Enhance the pure pixel signature using K-SVD as per the steps given below:

1.State the matrix $\omega_p = \{i | 1 \le i \le N, s_p(i) \ne 0\}$ to note down the non-zero indexes in s_p and define the maximum iteration time $t_{\max}$.

2.Compute E_p, the complete representation error matrix.

3.Confine E_p by selecting the columns vectors conforming to ω_p, and derive the E_p^R, the restricted error matrix.

4.Pick he modified pure pixel column a_p and update the corresponding coefficients vector using equation by Eq. (15) till the value of iterations reaches to $t_{\max}$.

$$\|X - AS\|_F^2 = \left\| X - \sum_{j=1}^{p} a_j s_T^j \right\|_F^2 = \left\| \left(X - \sum_{j \ne p} a_j s_T^j \right) - a_p s_T^p \right\|_F^2 = \left\| E_p - a_p s_T^p \right\|_F^2 \tag{15}$$

At the end, the endmember matrix, i.e., $A = [a_1, a_2, \cdots, a_p] \in R^{L \times P}$, is given as final outcome. As per the comparative evaluation done in [14], it is found that EESCA is more capable and robust to deal with the mixed data and noise respectively. Along with its advantages, it has also few downsides. EESCA neither considers the spatial features nor touch the number of endmember estimation problem.

In further portion, three SI based EE approaches (i.e., ACO, PSO and ABC) are discussed. Though these algorithms have capability to solve the problem of combinatorial optimization, their common drawback is that they are computationally expensive due to the random searching mechanism [26, 27].

2.8 Ant colony optimization (ACO)

The ant colony optimization (ACO) [15] was first employed for endmember extraction from hyperspectral imagery in 2011 and termed as ACOEE. In ACOEE the problem of decomposition of mixed pixels is transformed into a problem of optimization and forms a feasible solution space to estimate the practical implication of the objective function. To generate the solution space and the heuristic information for endmember extraction a directed and weighted graph G is constructed, where G corresponds to each pixel in the HSI data. The route generated by artificial ant (i.e., feasible solution) contains m unlike vertices in G, where, m represents the number of endmembers. On the arrival of artificial ant at the vertex v_i after $(t-1)$ times of moving, the probability of moving from v_i to vertex v_j is defined as [15],

$$p_{ij}^k(t) = \frac{\tau_{ij}^{\alpha}\eta_{ij}^{\beta}}{\sum_{j \in allowed_t} \tau_{ij}^{\alpha}\eta_{ij}^{\beta}}, \forall j \in allowed_t \tag{16}$$

where τ_{ij} shows the amount of pheromones in the edge $\langle v_i, v_j \rangle$ and $allowed_t$ indicates the set of pixels covered by ant from v_i (i.e., all pixels in the image except pixels that have been travelled by the ant) at time t. Parameters α and β shows relative significance of pheromones and visibilities in the selection of route respectively, where the pheromones concentration is initialized with similar value. On the basis of path travelled by an artificial ant (i.e., endmembers), a remixed image can be built by using abundance estimation. The root-mean-square error (RMSE) between the original and the remixed image is used as the objective function to calculate the endmembers set. In the kth iterative sequence, n ants creates dissimilar paths and pheromones in the edges updated using following equation,

$$\tau_{ij}^{k+1} = \rho\tau_{ij}^k + \Delta\tau_{ij}^k \tag{17}$$

where $\Delta\tau_{ij}^k$ is the pheromone increment and ρ is the factor of pheromone dissipation. If the smallest RMSE value in given repetition is f_k and its equivalent path is $path_k$, then $\Delta\tau_{ij}^k$ is,

$$\Delta\tau_{ij}^k = \begin{cases} Q/f_k \langle v_i, v_j \rangle \in path_k \\ 0 \langle v_i, v_j \rangle \notin path_k \end{cases}, \tag{18}$$

where Q represents constant which controls the rate of change of pheromones $\Delta\tau_{ij}^k$, within acceptable range. The algorithm breaks when it reaches at the similar optimal path in multiple serial repetitive sequences or to the maximum number [15, 27].

ACO EE outperforms the popular VCA and NFINDR EE algorithm and improves the endmember representation [15].

2.9 Particle swarm optimization (PSO)

The particle swarm optimization (PSO) [16] technique is a mostly used swarm intelligence technique to handle the problem of global optimization. Discrete PSO for endmember extraction (DPSO EE) was firstly adapted by Zhang et al. in 2011, where concept of PSO was utilized. In DPSO EE the combinatorial optimization problem is resolved by converting endmember extraction into feasible solution space and objective function, i.e., PSO searches for endmembers in the distinct feasible solution by outlining the particles locations and their velocities along with binary coding, and shows better endmember extraction results for HSI data than state of art methods [16, 28].

Linear spectral mixture model was used for spectral unmixing as shown in Eq. (19),

$$r_i = \sum_{j=1}^{m} e_j \alpha_{ij} + \varepsilon_i \tag{19}$$

where $\{r_i\}_{i=1}^{n}$ represents L bands and n pixels remote sensing image with r_i as column vector of spectrum of ith pixel, $\{e_i\}_{i=1}^{n}$ shows endmember set, ε_i is the random error and α_{ij} is the abundance of jth endmember in ith pixels. The least square method has been utilized for calculating area ratio of each endmember in a pixels, i.e., α_{ij}.

If the values of $\{r_i\}_{i=1}^{n}$ and $\{e_i\}_{i=1}^{n}$ are known then the image can remixed as, $r_i = \sum_{j=1}^{m} e_j \alpha_{ij},\ i = 1, 2, \ldots, n$. The root mean square error (rmse) between an original image and remixed image is calculated as [29],

$$rmse\left(\{r_i\}_{i=1}^{n}, \{e_i\}_{i=1}^{n}\right) = \frac{1}{n}\sum_{i=1}^{n}\sqrt{\frac{1}{L}}\left\| r_i - \widehat{r}_i \right\|_2^2 \tag{20}$$

If the value of rmse is less, then the result of endmember extraction is better. Therefore, endmember extraction can be defined as the combinatorial optimization problem as,

$$\begin{aligned} &\text{Min}\ \ rmse\left(\{r_i\}_{i=1}^{n}, E\right) \\ &\text{s.t.}\ \ E \in C\left(\{r_i\}_{i=1}^{n}, E, m\right) \end{aligned} \tag{21}$$

Here, $C\left(\{r_i\}_{i=1}^{n}, E, m\right)$ referred as, set of subsets of $\{r_i\}_{i=1}^{n}$ that contains m elements. $rmse\left(\{r_i\}_{i=1}^{n}, E\right)$ is the objective function and $C\left(\{r_i\}_{i=1}^{n}, E, m\right)$ is the feasible solution space.

The feasible solution space of the optimization problem is used to search particles by PSO. The adaptability function is obtained on the basis of constantly moving particles in the feasible space. In case of discrete feasible solution space D-PSO is improved to search particles in it on the basis of PSO. Moreover, the mapping relationship in the image and the feasible solution space can be given as [29],

$$\begin{aligned} &G : C\left(\{r_i\}_{i=1}^{n}, m\right) \rightarrow X_{n,m} \\ &E \mapsto (x_1, x_2, \ldots x_n). \end{aligned} \tag{22}$$

where $X_{n,m} = \left\{ (x_1, x_2, \ldots x_n) | x_i \in \{0, 1\}, \sum_{i=1}^{n} x_i = m \right\}$

From above formula, if a pixel r_i in $\{r_i\}_{i=1}^{n}$ is selected as an endmember, then value of x_i corresponding to r_i in x is 1; otherwise, the value is 0.

The DPSO EE shows better performance compared to VCA and NFINDR, but it has few limitations. The limitation includes several additional parameters that affect the performance of optimization and no assurance during about whether the location of selected particle is best or not [16].

2.10 Artificial bee colony (ABC)

The artificial bee colony (ABC) [17] algorithm is utilized for solving optimization problems by using the searching activity of bees in nature. The colony's search space is considered as feasible solution space for problem which is to be solved, where feasible solution is referred as a food source and quantity of nectar present in each source is referred as its fitness, which is linked with objective function value created by corresponding solution. Bees can be categorized into three types: employed bees, scout bees and onlooker bees. Xu Sun et al. used artificial bee colony for endmember extraction in 2015 [17, 27]. The step-wise workflow of ABC algorithm is given below:

a. The position of every employed bee is considered as position of a food source. The data about particular food source is recorded and fitness of respective food source is calculated. After finding with adjacent food sources, if fitness of newly found food source is better than previous one then update the former food source with new one otherwise continue to search.

b. If $X_i = (X_{i1}, X_{i2}, \ldots, X_{iM})^T$ represents the ith source food (i.e., location of employed bee) then adjacent searching can be defined as follows:

$$\varphi_{ij} = x_{ij} + \Delta\left(x_{ij} - x_{sj}\right) \tag{23}$$

where, Δ shows a random value in the range of $[-1, 1]$, $s \neq i, s \in \{1, 2, \ldots M\}$ and $j \in \{1, 2, .., M\}$ the greedy selection operator selects a food source with a better fitness value.

c. Each onlooker bee selects a food source on the basis of fitness acquired by the employed bees. It repeats similar policy of employed bee. The "follow probability" of an onlooker bee of selecting the jth food source is calculated as follows:

$$p_j = \frac{fit(x_j)}{\sum_{i=1}^{num} fit(x_i)} \tag{24}$$

where num shows the number of food sources which is greater than M and $fit(x_i)$ shows the fitness function of the ith food source and.

The fitness of a food source is determined using Eq. (25), where $f(x_i)$ is the objective function of the given optimization problem:

$$fit(x_i) = \frac{1}{f(x_i)} \tag{25}$$

The scout bee randomly reaches at a food source in the respective feasible solution space, converting itself into an employed bee to calculate the fitness value of the respective food source. Each random search can compute the best fitness position using the iterative process. If a food source has not been updated in a long time, then it is considered a global optimal source [17]. The ABC algorithm needs to face higher computational complexity for the images of larger sizes containing more number of endmembers.

3. Conclusion

The ultimate objective of using HSI data is to achieve the higher material classification or mapping accuracy, but it is adversely affected due the ground truth unavailability or spectral mixing problems. To overcome the same efficiently and enhance the HSI data classification or mapping accuracy, an image derived endmembers offers a precise and useful solution. To extract the endmembers from HSI data, several approaches based on either convex geometry or statistical information were tried over the past few years. Out of which, due to the higher volume of the data, the convex geometry based EE algorithms were found more effective and widely accepted and discussed in the literature. The popular algorithms comes under this criteria are PPI, VCA, SMACC, NFINDR, CCA and SGA. Apart from these techniques, some other statistical information and artificial intelligence based approaches are also well available in the literature, but unfortunately, none of the method provides the automated, data and application independent outcomes. Every method has its own limitations and challenges. These methods provides either inconsistent or misleading outcomes due to the random algorithm initialization nature, unclear criterion to define the values of additional input parameters and weakly defined algorithm stopping rules. In case of PPI, the algorithm is very sensitive to the value of iterations and threshold, where higher number of iterations results into higher computational complexity and unclear criteria to set the threshold value leads to the endmember selection problem. In case of VCA, SMACC and NFINDR, the basic requirement is that, the number of unique materials available into the image needs to be known primarily, which is hardly possible. To satisfy this requirement, some EE algorithms have opted the outcome of VD estimation (which gives the number of distinct signal sources available within the scene) as a reference to extract the number of endmembers. Further investigation of VD estimation techniques revealed that, these methods are also produce misleading results. More detailed and comparative analysis between various endmember extraction algorithms is performed time to time in [30–33]. Therefore, to overcome most of the challenges available in recent EE algorithm, there is a great need of a fully automated and robust endmember extraction algorithm. The efficient endmember extraction algorithm need to be data and application independent and must be able to find precise set of endmembers without any prior knowledge.

Acknowledgements

This work has been supported by DST, GOI, [under MRP No. BDID/01/23/2014-HSRS/35 (ALG-V)], UGS SAP II DRS Phase II, DST FIST and Department of Computer Science and Information technology, Dr. Babasaheb Ambedkar Marathwada University, Aurangabad-(MS), India.

Author details

Karbhari V. Kale*, Mahesh M. Solankar and Dhananjay B. Nalawade
Department of Computer Science and Information Technology, Dr. Babasaheb Ambedkar Marathwada University, Aurangabad, Maharashtra, India

*Address all correspondence to: kvkale91@gmail.com; kvkale.csit@bamu.ac.in

References

[1] Kale KV, Solankar MM, Nalawade DB, Dhumal RK, Gite HR. A research review on hyperspectral data processing and analysis algorithms. Proceedings of the National Academy of Sciences, India Section A: Physical Sciences. 2017;**87**(4):541-555

[2] Shippert P. Why use hyperspectral imagery? Photogrammetric Engineering and Remote Sensing. 2004;**70**(4): 377-396

[3] Landgrebe D. Hyperspectral image data analysis. IEEE Signal Processing Magazine. 2002;**19**(1):17-28

[4] Plaza A, Benediktsson JA, Boardman JW, Brazile J, Bruzzone L, Camps-Valls G, et al. Recent advances in techniques for hyperspectral image processing. Remote Sensing of Environment. 2009;**113**:S110-S122

[5] Mohan BK, Porwal A. Hyperspectral image processing and analysis. Current Science. 2015;**108**(25):833-841

[6] Chang CI. Hyperspectral Data Processing: Algorithm Design and Analysis. Hoboken, NJ: John Wiley & Sons; 2013

[7] Boardman JW, Kruse FA, Green RO. Mapping Target Signatures Via Partial Unmixing of AVIRIS Data

[8] Nascimento JM, Dias JM. Vertex component analysis: A fast algorithm to extract endmembers spectra from hyperspectral data. In: Iberian Conference on Pattern Recognition and Image Analysis; 4 June 2003; Berlin, Heidelberg: Springer; pp. 626-635

[9] Gruninger JH, Ratkowski AJ, Hoke ML. The sequential maximum angle convex cone (SMACC) endmember model. In: Algorithms and technologies for multispectral, hyperspectral, and ultraspectral imagery X. International Society for Optics and Photonics; 12 August 2004; Vol. 5425; pp. 1-15

[10] Winter ME. N-FINDR: An algorithm for fast autonomous spectral end-member determination in hyperspectral data. In: Imaging Spectrometry V. International Society for Optics and Photonics; 27 October 1999, Vol. 3753; pp. 266-276

[11] Chang CI, Wu CC, Liu W, Ouyang YC. A new growing method for simplex-based endmember extraction algorithm. IEEE Transactions on Geoscience and Remote Sensing. 2006; **44**(10):2804-2819

[12] Ifarraguerri A, Chang CI. Multispectral and hyperspectral image analysis with convex cones. IEEE Transactions on Geoscience and Remote Sensing. 1999;**37**(2):756-770

[13] Neville R. Automatic endmember extraction from hyperspectral data for mineral exploration. In: International Airborne Remote Sensing Conference and Exhibition, 4th/21st Canadian Symposium on Remote Sensing; Ottawa, Canada; 1999

[14] Wu K, Feng X, Xu H, Zhang Y. A novel endmember extraction method using sparse component analysis for hyperspectral remote sensing imagery. IEEE Access. 2018;**6**:75206-75215

[15] Zhang B, Sun X, Gao L, Yang L. Endmember extraction of hyperspectral remote sensing images based on the ant colony optimization (ACO) algorithm. IEEE Transactions on Geoscience and Remote Sensing. 2011;**49**(7):2635-2646

[16] Zhang B, Sun X, Gao L, Yang L. Endmember extraction of hyperspectral remote sensing images based on the discrete particle swarm optimization algorithm. IEEE Transactions on

Geoscience and Remote Sensing. 2011; **49**(11):4173-4176

[17] Sun X, Yang L, Zhang B, Gao L, Gao J. An endmember extraction method based on artificial bee colony algorithms for hyperspectral remote sensing images. Remote Sensing. 2015; **7**(12):16363-16383

[18] Zhu F, Wang Y, Fan B, Xiang S, Meng G, Pan C. Spectral unmixing via data-guided sparsity. IEEE Transactions on Image Processing. 2014;**23**(12): 5412-5427

[19] Plaza A, Chang CI. Fast implementation of pixel purity index algorithm. In: Algorithms and Technologies for Multispectral, Hyperspectral, and Ultraspectral Imagery XI. International Society for Optics and Photonics; 1 June 2005; Vol. 5806; pp. 307-318

[20] Chang CI, Plaza A. A fast iterative algorithm for implementation of pixel purity index. IEEE Geoscience and Remote Sensing Letters. 2006;**3**(1): 63-67

[21] Heylen R, Akhter MA, Scheunders P. A fast alternative for the pixel purity index algorithm. In: IGARSS 2015; IEEE. pp. 1781-1784

[22] Nascimento JM, Dias JM. Vertex component analysis: A fast algorithm to unmix hyperspectral data. IEEE Transactions on Geoscience and Remote Sensing. 2005;**43**(4):898-910

[23] Douglas WR, Laureano GT, Camilo CG. Comparison of VCA and GAEE algorithms for endmember extraction. In: 2018 IEEE Congress on Evolutionary Computation (CEC); 8 July 2018; IEEE. pp. 1-8

[24] Plaza A, Chang CI. An improved N-FINDR algorithm in implementation. In: Algorithms and Technologies for Multispectral, Hyperspectral, and Ultraspectral Imagery XI. International Society for Optics and Photonics; 1 June 2005; Vol. 5806; pp. 298-307

[25] Du Q, Raksuntorn N, Younan NH, King RL. Variants of N-FINDR algorithm for endmember extraction. In: Image and Signal Processing for Remote Sensing XIV. International Society for Optics and Photonics; 10 October 2008; Vol. 7109; p. 71090G.

[26] Dorigo M, Birattari M. Swarm intelligence. Scholarpedia. 2007;**2**(9): 1462

[27] Su Y, Sun X, Gao L, Li J, Zhang B. Improved discrete swarm intelligence algorithms for endmember extraction from hyperspectral remote sensing images. Journal of Applied Remote Sensing. 2016;**10**(4):045018

[28] Xu M, Zhang L, Du B, Zhang L, Fan Y, Song D. A mutation operator accelerated quantum-behaved particle swarm optimization algorithm for hyperspectral endmember extraction. Remote Sensing. 2017;**9**(3):197

[29] Tembhurne OW, Shrimankar D. EEA-PSO: Endmember extraction using advance particle swarm optimization. In: 2018 4th International Conference on Recent Advances in Information Technology (RAIT); 15 March 2018; IEEE. pp. 1-8

[30] Plaza A, Martínez P, Pérez R, Plaza J. A quantitative and comparative analysis of endmember extraction algorithms from hyperspectral data. IEEE Transactions on Geoscience and Remote Sensing. 2004;**42**(3): 650-663

[31] Plaza A, Sánchez-Testal JJ, Plaza J, Valencia D. An experimental evaluation of endmember generation algorithms. In: Chemical and Biological Standoff Detection III. Vol. 5995. International Society for Optics and Photonics; 2005. p. 599501

[32] Plaza J, Hendrix EM, García I, Martín G, Plaza A. On endmember identification in hyperspectral images without pure pixels: A comparison of algorithms. Journal of Mathematical Imaging and Vision. 2012;**42**(2–3): 163-175

[33] Solankar MM, Gite HR, Dhumal RK, Surase RR, Nalawade D, Kale KV. Recent advances and challenges in automatic hyperspectral endmember extraction. In: Proceedings of 2nd International Conference on Communication, Computing and Networking 2019; Singapore: Springer; pp. 445-455

Chapter 2

Hyperspectral Image Classification

Rajesh Gogineni and Ashvini Chaturvedi

Abstract

Hyperspectral image (HSI) classification is a phenomenal mechanism to analyze diversified land cover in remotely sensed hyperspectral images. In the field of remote sensing, HSI classification has been an established research topic, and herein, the inherent primary challenges are (i) curse of dimensionality and (ii) insufficient samples pool during training. Given a set of observations with known class labels, the basic goal of hyperspectral image classification is to assign a class label to each pixel. This chapter discusses the recent progress in the classification of HS images in the aspects of Kernel-based methods, supervised and unsupervised classifiers, classification based on sparse representation, and spectral-spatial classification. Further, the classification methods based on machine learning and the future directions are discussed.

Keywords: hyperspectral imaging, classification, supervised and unsupervised classification, machine learning

1. Introduction

The technological progression in optical sensors over the last few decades provides enormous amount of information in terms of attaining requisite spatial, spectral and temporal resolutions. Especially, the generous spectral information comprises of hyperspectral images (HSIs) establishes new application domains and poses new technological challenges in data analysis [1]. With the available high spectral resolution, subtle objects and materials can be extracted by hyperspectral imaging sensors with very narrow diagnostic spectral bands for the variety of purposes such as detection, urban planning [2], agriculture [3], identification, surveillance [4], and quantification [5, 6]. HSIs allow the characterization of objects of interest (e.g., land cover classes) with unprecedented accuracy, and keep inventories up to date. Improvements in spectral resolution have called for advances in signal processing and exploitation algorithms.

Hyperspectral image is a 3D data cube, which contains two-dimensional spatial information (image feature) and one-dimensional spectral information (spectral-bands). Especially, the spectral bands occupy very fine wavelengths, while the image features such as Land cover features and shape features disclose the disparity and association among adjacent pixels from different directions at a confident wavelength.

In the remote sensing community, the term classification is used to denote the process that assigns individual pixels to a set of classes. The output of the classification step is known as the classification map. With respect to the availability of training samples, classification approaches can be split into two categories, i.e., supervised and unsupervised classifiers. Supervised approaches classify input data for each class using a set of representative samples known as training samples. Hyperspectral (HS) image classification always suffers from varieties of artifacts, such as high dimensionality, limited or unbalanced training samples [7], spectral variability, and mixing pixels. The Hughes phenomenon is a common problem in the supervised classification process [8]. The power of classification increases with the increase of available training samples. The limited availability of training samples decreases the classification performance with the increase of feature dimension. This effect is famously termed as "Hughes phenomenon" [9]. It is well known that increasing data dimensionality and high redundancy between features might cause problems during data analysis. There are many significant challenges that need to be addressed when performing hyperspectral image classification. Primarily, supervised classification faces challenge about the imbalance between high dimensionality and incomplete accessibility of training samples or the presence of mixed pixels in the data [10]. Further, it is desirable to integrate the essential spatial as well as spectral information so as to combine the complementary features that stem from source images [11]. A considerable amount of literature has been published with regard to overcoming these challenges, and performing hyperspectral image classification effectively.

Hyperspectral image classification could attract scientific community which aims at assigning a pixel (or a spectrum) to one of a certain set of predefined classes. Maximum likelihood (ML) methods, neural networks architectures [12], support vector machine (SVM) [13], Bayesian approach [14] as well as kernel methods [15] are the prominent methods which have been investigated in recent years for the identification or classification of hyperspectral data.

Based on the usage of training sample, image classification task is categorized as supervised, unsupervised and semi-supervised hyperspectral image classification.

2. Unsupervised classification

The paramount challenge for HSI classification is the curse of dimensionality which is also termed as Hughes phenomenon. To confront with this difficulty, feature extraction methods are used to reduce the dimensionality by selecting the prominent features. In unsupervised methods, the algorithm or method automatically groups pixels with similar spectral characteristics (means, standard deviations, etc.) into unique clusters according to some statistically determined criteria. Further, unsupervised classification methods do not require any prior knowledge to train the data. The familiar unsupervised methods are principal component analysis (PCA) [16] and independent component analysis (ICA) [17].

2.1 Principal component analysis

It is the most widely used technique for dimensionality reduction. In comparative sense, appreciable reduction in the number of variables is possible while retaining most of the information contained by the original dataset. The substantial correlation between the hyperspectral bands is the basis for PCA. The analysis attempts to eliminate the correlation between the bands and further determines the optimum linear combination of the original bands accounting for the variation of pixel values in an image [18].

The mathematical principle of PCA relies upon the eigen value decomposition of covariance matrix of HSI bands. The pixels of hyperspectral data are arranged as a vector having its size same as the number of bands. $X_i = [x_1, x_2, \ldots \ldots x_N]^T$, where N is the number of HS bands. The mean of all the pixel vectors is calculated as:

$$m = \frac{1}{M}\sum_{i=1}^{M} [x_1\ x_2 \ldots x_N]_i^T \tag{1}$$

where M = p ⋆ q is the number of pixel vectors for a HS image of "p" rows and "q" columns. The covariance matrix is determined as:

$$C = \frac{1}{M}\sum_{i=1}^{M} (X_i - m)(X_i - m)^T \tag{2}$$

The covariance matrix can also be written as:

$$C = ADA^T \tag{3}$$

D is the diagonal matrix composed of eigen values $\{\lambda_1, \ldots \ldots \lambda_N\}$ of C and A is the orthogonal matrix with the corresponding eigen vectors (each of size N) as columns. The linear transformation $y_i = A^T X_i, i = 1, 2. \ldots M$, is adapted to achieve the modified pixel vectors which are the PCA transformed bands of original images. The first K rows of the matrix A^T are selected such that, the rows are the eigen vectors corresponding to the eigen values arranged in a descending order. The selected K rows are multiplied with the pixel vector X_i to yield the PCA bands composed of most of the information contained in the HS bands.

In hypespectral data, most of the elements are covered by the sensors with high spectral resolution which cannot be well described by the second order characteristics. Hence, PCA is not an effective tool for HS image classification since it deals with only second-order statistics.

2.2 Independent component analysis (ICA)

Independent component analysis successfully executes the independence of the components with higher-order statistics, and is relatively more suitable to encounter high dimensionality of HS images. ICA is an attractive tool for dimensionality reduction, feature extraction, blind source separation, etc., as well as to preserve the information which cannot be retrieved using second order statistics [19, 20].

Let us consider a mixture of random variables $x_1, x_2, \ldots x_N$, where each $x_i \in R^d$. These random variables are defined as a linear combination of another random variables $p_1, p_2, \ldots, p_N$, where each $p_i \in R^n$. In such scenario, the mixing model can be mathematically written as,

$$X = AP \tag{4}$$

where $X = [x_1, x_2, \ldots, x_N]$ is the observed vector, $P = [p_1, p_2, \ldots, p_N]$ is the unknown source, A is the mixing matrix, "n" denotes the number of unknown sources and "d" represents the number of observations made. In order to find the independent components, the unmixing matrix W is to be estimated (inverse of A). The independent components are obtained using Eq. (5).

$$ICA(X) = P = A^{-1}X = WX \tag{5}$$

If $X \in R^{d \times N}$ is considered as the hyperspectral image,

$$P_{n \times N} = W_{n \times d} X_{d \times N} \tag{6}$$

where N is the number of pixels in each band, d represents the number of spectral bands and n gives the number of sources or materials present in the image. The estimation of the ICA model is conceivable, only if the following presumptions and limitations are fulfilled: (i) Sources should be statistically independent (ii) Independent components should possess non Gaussian distribution (iii) Matrix A should be a square and full rank matrix.

3. Supervised classification

The supervised classification takes the advantage of rich spectral information and has explored many applications including urban development [21], the monitoring of land changes [22], target detection [23], and resource management [24]. In supervised classification only labeled data is used to train the classifier. A large number of supervised classification methods have been discussed in the literature, some of the prominent methods are maximum likelihood (ML), nearest neighbor classifier, decision trees, random forest, support vector machines (SVMs), etc. **Figure 1** shows the conventional steps of supervised classification of HSIs.

3.1 ML classifier

The ML classifier assumes that the statistics for each class in each band are normally distributed and estimates the probability that a given pixel belongs to a certain specific class [25]. Unless a probability threshold is selected, all pixels are classified. Each pixel is assigned to a particular class that manifests the maximum probability. If the estimated maximum probability is smaller than a threshold, the pixel remains unclassified. The following discriminant functions for each pixel in the image are implemented in ML classification.

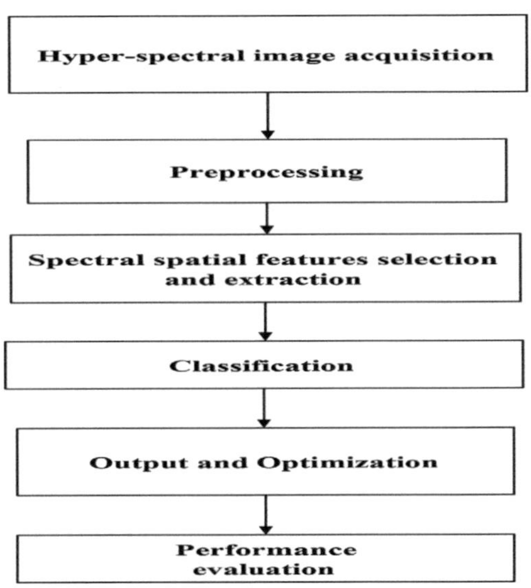

Figure 1.
Flowchart of HSI supervised classification.

$$g_i(x) = \ln\ p(w_i) - \frac{1}{2}\ln|\sigma_i| - \frac{1}{2}(x - m_i)^t\sigma_i^{-1}(x - m_i) \tag{7}$$

where i = class; x = n-dimensional data (where n represents the number of bands); $p(w_i)$ = probability that class w_i occurs in the image and is assumed the same for all classes; $|\sigma_i|$ = determinant of the covariance matrix of the data in class w_i; $|\sigma_i|^{-1}$ = its inverse matrix; and m_i = mean vector.

Implementation of the ML classification involves the estimation of class mean vectors and covariance matrices using training pattern chosen from known examples of each particular class [26]. It usually acquires higher classification accuracy compared to other traditional classification approaches. It assumes that each band is normally distributed and the chosen training samples are comprised of exhaustively defined set of classes. For hyperspectral data with tens of hundreds of spectral bands, discrimination of land cover classes is not an easy task, whereas, the classification accuracy of ML classifier is based on the accurate selection of the training samples. Thus, for the hyperspectral imagery with poorly represented labeled training samples, it is preferable to adapt an alternative to the standard multiclass classifier.

3.2 *k*-nearest-neighbor (kNN) classifier

*k*NN is one of the widely used simplest classifier, and has been applied for HSI classification [27, 28].

kNN method operates on majority voting rule, presumes that all the neighbors make equal contributions to the classification of the testing point. Another important feature of kNN classifier is Euclidian is used as distance metric, which assumes the data is homogeneous.

Let $\mathbf{X} = [\mathbf{x}_1, \dots, \mathbf{x}_N]$ be the N-point training data, with d as the dimension of each point. $\mathbf{X}_i = [\mathbf{x}_{i1}, \dots, \mathbf{x}_{ik}]$ be the k nearest neighbors of $\mathbf{x}_i$. The testing data (N_t points) is denoted as X_t with x_0 is a random testing point. The k nearest neighbors from the testing data with labels $[l_1, l_2 \dots l_k]$ is indicated as $\mathbf{X}_0 = [\mathbf{x}_{01}, \dots, \mathbf{x}_{0k}]$. Let assume that $[\Omega] = [\Omega_1, \dots, \Omega_C]$ are the "C" classes in the data.

The kNN classifier finds the k nearest neighbors of a testing point in the training data and assigns the testing point to the most frequently occurring class of its k neighbors. The classification of $\mathbf{x}_0$ by majority voting rule is exercised using the following expression:

$$j^* = arg \max_{j=1, \dots, C} \sum_{i=1}^{k} \delta(l_i, j) \tag{8}$$

where δ is the Kronecker delta.

A distance metric learned from the given training data is used to enhance the accuracy of *k*NN classifier.

$$dis(\mathbf{x}_i, \mathbf{x}_j) = \left\|\mathbf{T}(\mathbf{x}_i - \mathbf{x}_j)\right\|^2 \tag{9}$$

T denotes a linear transformation.

The decision rule of *k*NN can be modified by assigning different weights to the neighbors. Further, the testing point is assigned to the class for which the sum of weights chosen for the neighbors is largest.

$$j^* = arg \max_{j=1, \dots, C} \sum_{i=1}^{k} w_i\delta(l_i, j) \tag{10}$$

It is also referred as decision rule for weighted *k*NN (W*k*NN), where w_i is the weight of x_{0i}.

3.3 Spectral angle mapper (SAM)

SAM is a supervised classification technique for HSIC [29]. SAM classifier admits very quick classification using the spectral angle information of HSI data.

The reference spectra are usually determined from the field measurements or from the image data, is used to measure the spectral angle. The spectral angle is a n-dimensional vector between image and reference spectra. Smaller the angles between two spectrums, higher the similarity and vice versa. The classification approach using SAM is described in **Figure 2**.

This technique is comparatively insensitive to illumination and albedo effects when reflectance data is used for analysis. The spectral angle can be calculated as follows:

$$\theta = \cos^{-1}\left(\frac{\sum_{i=1}^{N} T_i R_i}{\sqrt{\sum_{i=1}^{N} T_i^2}\sqrt{\sum_{i=1}^{N} R_i^2}}\right) \tag{11}$$

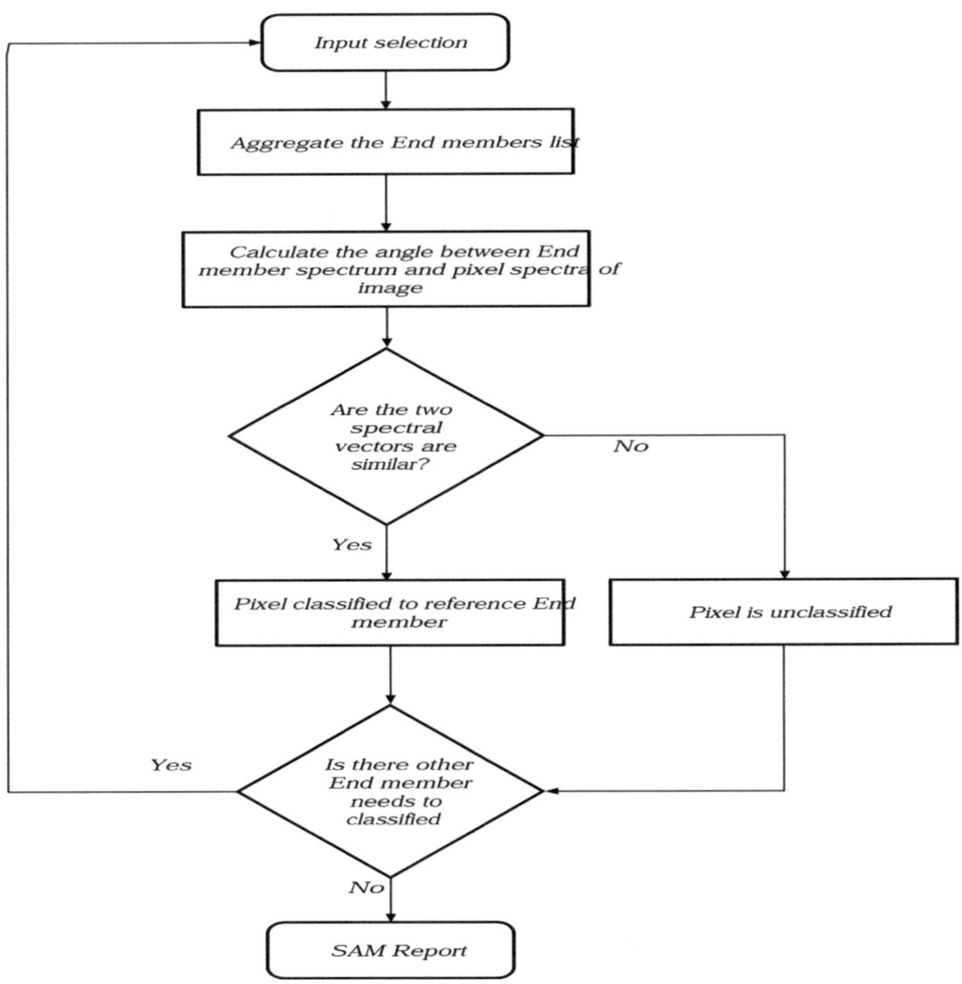

Figure 2.
SAM classification approach.

3.4 Support vector machine (SVM)

SVM is typically a linear classifier associative with kernel functions and optimization theory and is prominent for HSI classification [13, 30, 31]. SVM outperforms the conventional supervised classification methods particularly in prevailing conditions like increased number of spectral bands and the limited availability of training samples [32–34].

3.4.1 Linear SVM: Linearly separable case

Let $x_i \in \mathbb{R}^d$, $(i = 1, 2 \dots N)$ be the set of training vectors, and a target $y_i \in \{-1, +1\}$ is corresponding to each vector x_i. The problem is treated as a binary classification and the two classes are linearly separable. Hence, at least one hyperplane must exist to separate the two classes without errors. The discriminant function associated with hyperplane can be defined as:

$$f(x) = wx + b \tag{12}$$

where $w \in \mathbb{R}^d$ is a vector normal to hyperplane, $b \in \mathbb{R}$ is a bias. w and b must satisfy the following condition to estimate such a hyperplane,

$$y_i(w.x_i + b) > 0, \quad \text{for } i = 1, 2 \dots N \tag{13}$$

The optimal hyperplane can be estimated by solving the following convex problem.

$$\min \frac{1}{2}\|w\|^2 \text{ s.t } y_i(w.x_i + b) \geq 1, \text{for } i = 1, 2 \dots N \tag{14}$$

3.4.2 Linearly nonseparable case

For practical data classification problem, the linearly separable condition may not be true in different conditions. To solve the classification problem of nonseparable data, hyperplane separation has been generalized. A cost function is formulated comprising two conditions: margin maximization (as in the case of linearly separable data) and error minimization (to penalize the wrongly classified samples).

$$\psi(w, \xi) = \frac{1}{2}\|w\|^2 + C\sum_{i=1}^{N} \xi_i \tag{15}$$

Where, ξ_i are slack variables derived to account for the nonseparability of data and C is a regularization parameter. The larger the C value, the higher the penalty associated with misclassified sample.

The minimization of the cost function defined in Eq. (15) is subject to the following conditions:

$$y_i(w.x_i + b) \geq 1 - \xi_i, i = 1, 2. \dots N. \tag{16}$$

$$\xi_i \geq 0, \quad i = 1, 2. \dots N. \tag{17}$$

For nonseparable data, two types of support vectors coexist: (1) margin support vectors that lie on the hyperplane margin and (2) nonmargin support vectors that fall on the "wrong" side of this margin [13].

3.4.3 Nonlinear SVM -kernel method

The effective discriminant function to solve the nonlinear classification problem can be expressed as:

$$f(x) = \sum_{i \in S} \alpha_i y_i K(x_i, x) + b \tag{18}$$

A common example of kernel type that fulfills Mercer's condition is the Gaussian radial basis function:

$$K(x_i, x) = \exp\left(-\gamma \|x_i - x\|^2\right) \tag{19}$$

where, γ is a parameter that is inversely proportional to width of the Gaussian kernel. The more details about kernel functions for this case can be referred in [35].

4. Random forest classifier

A random forest (RF) is a group of tree-based classifiers where each tree is trained with a bootstrapped set of training data. The data to be classified is applied as an input to each tree in the forest. The classification given by each tree is known as a "vote" for that class. In the classification, the forest chooses the class having the most votes (over all the trees in the forest). In RF classification a split is determined by searching across a random subset of variables at each node [36, 37].

The Random forest classifier (RFC) features two main characteristics: relatively high accuracy and the speed of processing. However, the correlation/independence of trees can affect the accuracy of final land cover map. The primitive components of Random Forest are explored as:

4.1 CART-like trees

Classification and regression tree (CART), a binary tree in which splits are resolved by the variables obtained from the strong change in impurity or minimum impurity ($\hat{i}(t)$),

$$\hat{i}(t) = \sum_{i \neq j} \hat{P}(x_i \mid t)\hat{P}(x_j \mid t) \tag{20}$$

where $\hat{P}(x_i|t)$ is the estimated probability of sample $x_i \in$ class i. The definite classification takes place during training process. Either the impurity is zero or all the splits result in only one node then the growth of the tree terminates.

4.2 Binary hierarchy classifier (BHC)

In contrary to CART, the split on each node in BHC is based on classes. The optimal split at each node is based on class separability and further the splits are pure.

Let us consider a single meta-class case, which split into two into 2 meta-classes and so on, until the true classes are realized in the leaves, while simultaneously computing the Fisher discriminant and projection.

Let μ_γ, and σ_γ, $\gamma \in \{y, \beta\}$ are the estimated mean vector and co-variance matrix of the meta class w_γ, then the data projected using w:

$$w = W^{-1}(\mu_\alpha - \mu_\beta) \tag{21}$$

The inverse of class covariance matrix W

$$W = P(\omega_\alpha)\sigma_\alpha + P(\omega_\beta)\sigma_\beta \tag{22}$$

P() is a prior probability. The discriminant $\mathcal{T}(W)$ can be maximized as:

$$\mathcal{T}(w) = \frac{w^T B w}{w^T W w} \tag{23}$$

Where, B is the covariance matrix between classes.

$$B = (\mu_\alpha - \mu_\beta)(\mu_\alpha - \mu_\beta)^T. \tag{24}$$

Like the CART trees, the BHC trees can be combined as a forest (RF-BHC) to realize an ensemble of classifiers, where the best splits on classes are performed on a subset of the features in the data to diversify individual trees and/or to stabilize the W.

5. Spatial-spectral classification

The pixel-wise classification methods incur some difficulties: Discriminating the classes is very difficult due to less interclass spectral variability. If interclass variability is high, it is very hard to determine a given class. The pixel-wise classification capability can be enhanced by the exploration of additional information called spatial dependency. The classification performance can be improved by incorporating spatial information into HSIC. This rationale motivates the study of spatial-spectral classification methodologies [38]. The spatial dependency system for spectral-spatial-based classification is depicted in **Figure 3**. The spatial dependency (primary information for spatial-spectral classification techniques) is carried by two identities called pixel and associated label. The correlation among spatially related pixels is spatial dependency, hence spatially related pixels are termed as neighboring pixels. The spatial dependency is associated with (i) Pixel dependency indicates the correlation of neighboring pixels and (ii) Label dependency indicates the correlation of labels of neighboring pixels. Distinct approaches of spatial-spectral classification are as follows [39]:

i. Structural filtering: The spatial information from a region of the hyperspectral data is extracted by evaluating the metrics like mean and standard deviation of neighboring pixels over a window. The relevant methods include spectral-spatial wavelet features [40], Gabor features [41], Wiener filtering [42], etc.

ii. Morphological profile (MP): mathematical morphology (MM) intent to investigate spatial relationships between pixels using a set of known shape and size which is called the structuring element (SE). Dilation and erosion are the two elemental MM operations used for nonlinear image processing. The concept of extracting the information regarding contrast and size of the structures present in an image is termed as granulometry. The morphological profile (MP) of size n has been defined as the composition of a granulometry of size n built with opening by reconstruction and a (anti)granulometry of size n built with closing by reconstruction [43].

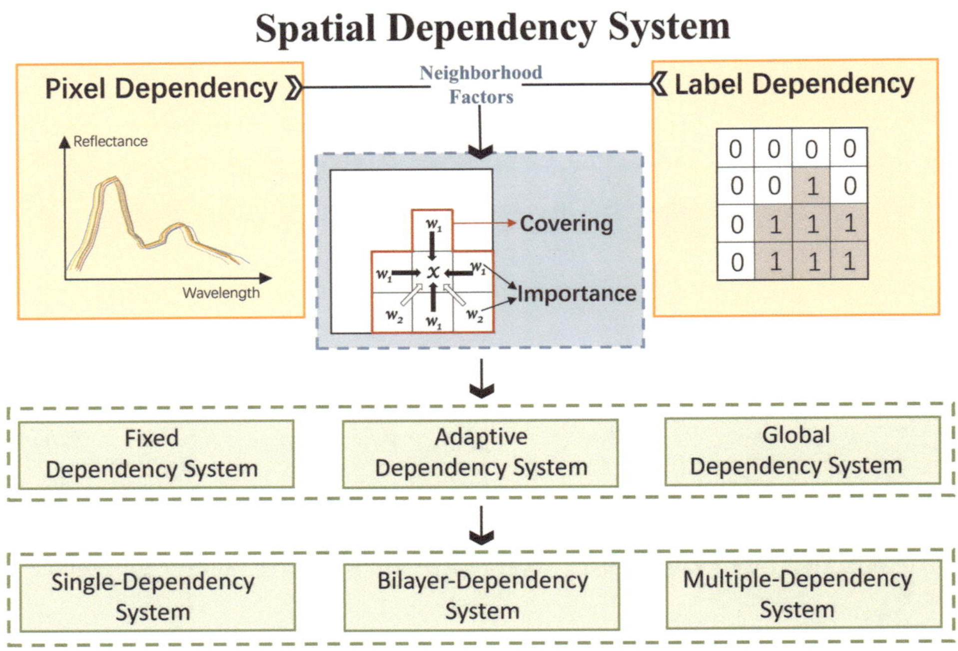

Figure 3.
Spatial dependency system in spectral-spatial classification.

$$MP^{(n)}(I) = \left[\phi_r^{(n)}(I), \ldots, \phi_r^{(1)}(I), I, \gamma_r^{(1)}(I), \ldots, \gamma_r^{(n)}(I)\right] \quad (25)$$

From a single panchromatic image, the MP results in a (2n + 1)-band image. However, for hyperspectral images the direct construction of the MP is not straightforward, because of the lack of ordering relation between vector. In order to overcome this shortcoming, several approaches have been considered [44].

i. Random field: random field-based methods have been studied broadly for HSI classification. Markov random fields (MRFs) and conditional random fields (CRFs) are two major variants of RF-based classification methods. CRF methods adapt conditional probability for labeling the data and attain favorable performance by utilizing the optimal spatial information; whereas, MRF-based techniques achieve substantial reduction in computational complexity by estimating class parameters independently from field parameters. The basic formulation of random fields as follows:

Let $\mathcal{S} = \{1, \ldots\ldots, n\}$ denote a set of integers indexing the n pixels of a hyperspectral image. A conditional probability $P(y/x)$ (a posteriori) is defined with $x = \{x_1, x_2, \ldots\ldots x_n\} \in R^{d\times n}$ denotes d-dimensional feature vectors composes a hyperspectral image and $y = \{y_1, y_2 \ldots y_n\}$ is an image of lables. The a posteriori probability can be expressed as:

$$p(\mathbf{y}/\mathbf{x}) = \frac{1}{Z(\omega, \mathbf{x})} \exp\left(\sum_{i\in\mathcal{S}} \log p(y_i|\mathbf{x}_i, \omega) + \mu \sum_{(i,j)\in C} \delta(y_i - y_j)\right) \quad (26)$$

The normalizing facor $Z(\omega, x)$, also known as partition function is defined as:

$$Z(\omega, \mathbf{x}) = \sum_{\mathbf{y}} \exp\left(\sum_{i \in S} \log p(y_i|\mathbf{x}_i, \omega) + \mu \sum_{(i,j) \in C} \delta\left(y_i - y_j\right)\right) \tag{27}$$

where, $p(y_i|\mathbf{x}_i, \omega)$ =the class probability given by the learning parameter ω.
μ = parameter controlling the degree of smoothness on the image of labels.
$\delta(y)$ = unit impulse function and C is a set of cliques.

The CRFs not only avoids label bias problem but also its conditional nature motivates the relaxation of independence assumptions. Recently, Distributed random Forest (DRF) have gained interest for HSIC [45] owing to its inherent merit.

The salient features of DRF are (1) the relaxation of conditional independence of the observed data. (2) the exploitation of probabilistic discriminative models instead of the generative MRFs. and (3) the simultaneous estimation of all DRF parameters from the training data.

6. Sparse-representation (SR)-based classification

The role of SR theory has become prevalent in almost all the image processing applications. The SR theory presumes that the training samples can be represented as a linear combination of smallest possible number of atoms (columns) of an over-complete dictionary.

The test sample x_i can be represented as $x_i = D\alpha + \epsilon$. where, $D \in R^{n \times k}$ is a dictionary with n samples of k dimensions and the sparse coefficients vector α can be determined by solving the following optimization problem.

$$\hat{\boldsymbol{\alpha}} = arg \min \|\boldsymbol{\alpha}\|_0 \quad \text{s.t.} \|\mathbf{x}_i - \mathbf{D}\boldsymbol{\alpha}\|_2 \le \epsilon \tag{28}$$

The term $\|.\|_0$ is l_0 norm that counts the number of nonzero entries. The optimization problem in Eq. (28) can be solved with greedy pursuit algorithms [46], in which the l_0 norm is replaced with the l_1 norm.

For HSIC, the Eq. (28) can be replaced as:

$$\min_{\boldsymbol{\alpha}} \frac{1}{2} \|\mathbf{x}_i - \mathbf{D}\boldsymbol{\alpha}\|_2^2 + \tau \|\boldsymbol{\alpha}\|_1, \quad \boldsymbol{\alpha} \ge \mathbf{0}. \tag{29}$$

where, the parameter τ is a Lagrange multiplier that balances the tradeoff between the reconstruction error and the sparse solution: $\tau \to 0$ when $\epsilon \to 0$.

In order to incorporate the spatial information a spatial weight is added and the modified SR model for HSIC is formulated as:

$$\min_{\boldsymbol{\alpha}} \frac{1}{2} \|\mathbf{x}_i - \mathbf{D}\boldsymbol{\alpha}\|_2^2 + \tau \|\mathbf{W}\boldsymbol{\alpha}\|_1, \quad \boldsymbol{\alpha} \ge \mathbf{0} \tag{30}$$

The choice of a spatial weight matrix W, yields different classification strategies for HSIs namely neighboring pixels [47], neighboring filtering [38], histogram-based [47], spatial information based on super pixels [48], etc.

The class labels can be implied on the basis of the following formulation:

$$\hat{\text{class}}(\mathbf{x}_i) = arg \min_{j \in \{1, \dots, c\}} \|\mathbf{x}_i - \mathbf{D}_j \boldsymbol{\alpha}_j\|_2. \tag{31}$$

A sparsity-based algorithm to improve the classification performance is proposed in [49]. The principle depends on the sparse representation of a hyperspectral pixel by a linear combination of a few training samples from a structured dictionary. The sparse vector is recovered by solving a sparsity-constrained optimization problem, and it can directly determine the class label of the test sample. Zhang et al. [50] proposed a nonlocal weighted joint sparse representation (NLW-JSRC) to further improve the classification accuracy. The method enforced a weight matrix on the pixels of a patch in order to discard the invalid pixels whose class was different from that of the central pixel. A few of the recent investigations [51–53] approved that a compact and discriminative dictionary learned from the training samples can significantly reduce the computational complexity.

6.1 Segmentation-based methodologies

The segmentation process is performed after spectral-based classification in some of HSIC techniques. The extraction and classification of homogeneous objects is presented in [54] is the first classifier that used spatial postprocessing. The comprehensive survey of other methodologies of this category is presented in [43].

7. Deep learning (DL)

Deep learning involves a class of models which try to hierarchically learn deep features of input data with very deep neural networks, typically deeper than three layers. The network is first layer-wise initialized via unsupervised training and subsequently, tuned in a supervised manner. In this scheme, high level features are learned from low level ones, whereas, the proper features can be formulated for pattern classification towards the end. Deep models can potentially lead to progressively more abstract and complex features at higher layers, and more abstract features are generally invariant to the most local changes experienced by the input data.

7.1 Deep learning for HSI classification

The DL theory presents a dynamic way for unsupervised feature learning using very large raw image dataset. Unlike the traditional classification techniques, DL-based techniques can represent and organize multiple levels of information to express complex relationships between data.

Deep Learning (DL) is a sort of more complex architecture simulating human brains, based on neural networks begins to apply hyperspectral image classification [55]. The deep learning models for HSIC usually consists of three layers, to extract the more complex characteristics layer by layer. (i) Input data (ii) Deep layer construction (iii) Classification [56]. The notable methodologies include deep belief network (DBN) [57], stacked auto encoder (SAE) [58], and convolutional neural network (CNN) [59].

Deep belief networks (DBNs) [60] are an important development in DL research and train one layer at a time in an unsupervised manner by restricted Boltzmann machines (RBMs) [61]. The DBNs admit unsupervised pretraining over unlabeled samples at first and then a supervised fine-tuning over labeled samples. Since the pretrained DBN captures the useful information from the unlabeled samples, the fine-tuning with the pretrained DBN performes well over small number of labeled samples [57, 62]. The simple structure of DBN is presented in **Figure 4**.

The conventional training of DBN incur two problems; The first is coadaptation of latent factors [63, 64]. This activity is described as several latent factors tend to behave very similarly. This phenomenon implies that the model parameters

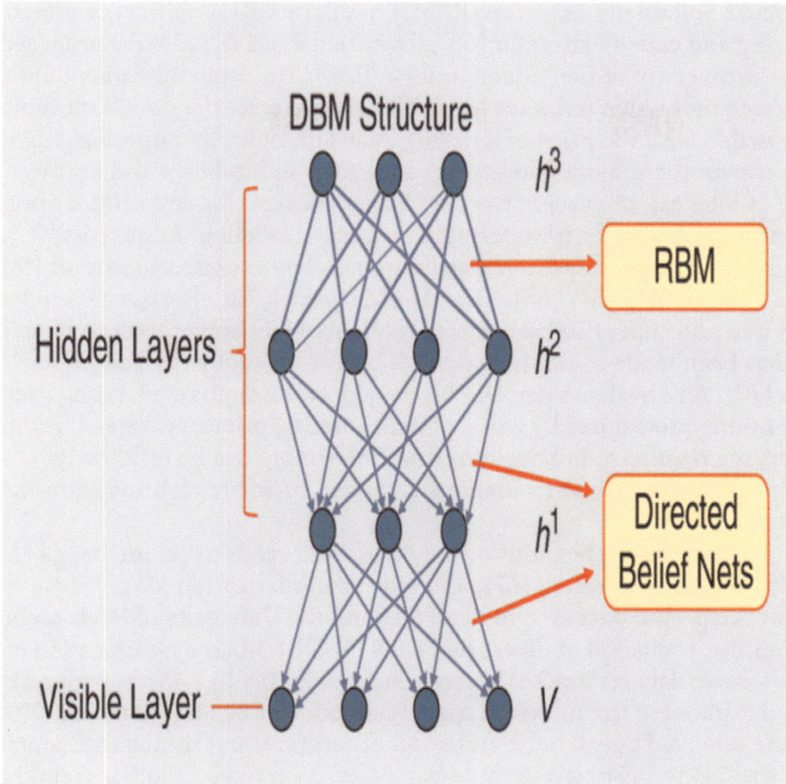

Figure 4.
The simple structure of the standard DBN. (RBM- Restricted Boltzmann Machine).

corresponding to the latent factors might be very similar. These similar latent factors make most of the computations to be performed redundantly and also decrease DBN's description ability. The second is the set of many "dead" (never responding) or "potential over-tolerant" (always responding) latent factors (neurons) in the DBN learned with the usual sparsity promoting priors [65]. The "dead" or "potential over-tolerant" latent factors directly correspond to the decrease of the model's description sources. These problems reduce the DBN's description ability as well as the classification performance. The first problem is solved by trying to perform the latent factors diversely. The "dead" and "potential over-tolerant" latent factors (neurons) are related to the sparsity and selectivity of activations of visual neurons and the selectivity and sparsity are just two epiphenomena of the diversity of receptive fields. Hence, both the problems can be solved together by diversifying the DBN models.

The classification performance enhancement through the diversification of latent factors of a given model has became attractive topic in recent years [66–68]. The determinantal point process (DPP) is used as a prior for probabilistic latent variable models in [68]. Probabilistic latent variable models are one of the vital elements of machine learning. The determinantal point process enables a modeler to specify a notion of similarity on the space of interest, which in this case is a space of possible latent distributions, via a positive definite kernel. The DPP then assigns probabilities to particular configurations of these distributions according to the determinant of the Gram matrix. This construction naturally leads to a generative latent variable model in which diverse sets of latent parameters are preferred over redundant sets.

Restricted Boltzmann Machine (RBM)s has demonstrate immense effectiveness in clustering and classification. In [69], divesified RBM (DRDM) is proposed to enhance the diversity of the hidden units in RBM. To combat the phenomenon that many redundant hidden units are learned to characterize the dominant topics as best as possible with the price of ignoring long-tail topics by imposing a diversity regularizer over these hidden units to reduce their redundancy and improve their coverage of long-tail topics. First-order Hidden Markov Models (HMM) provides a fundamental approach for unsupervised sequential labeling. A diversity-encouraging prior over transition distributions is incorporated to extend HMM to diversified HMM (dHMM) [66]. The dHMM shows great effectiveness in both the unsupervised and supervised settings of sequential labeling problems. A successful attempt has been made to improve the HSI classification by diversifying a deep model in [70]. A new diversified DBN is developed through regularizing pretraining and fine-tuning procedures by a diversity promoting prior over latent factors. Moreover, the regularized pretraining and fine-tuning can be efficiently implemented through usual recursive greedy and back-propagation learning framework.

The conventional applications of the diversified models include image classification [69], image restoration [67], and video summarization [71].

Two hyperspectral data sets, Indian Pines and the University of Pavia scenes are selected for the evaluation of diversified DBN (D-DBN)-based classification method. The Indian Pines data set has 220 spectral channels in 0.4 to 2.45 μm region of the visible and infrared spectrum with a spatial resolution of 20 m × 20 m. The 20 spectral bands were removed due to noise and water absorption, and the data set contains 200 bands of size 145 × 145 pixels. A three-band false color image and the ground truth data are presented in **Figure 5**. The University of Pavia data set with a spectral coverage ranging from 0.43 to 0.86 μm is presented in **Figure 6**. The image contains 610 × 340 pixels and 115 bands. After removing 12 bands due to noise and water absorption, the image contains 103 bands with a spatial resolution as 1.3 m × 1.3 m.

The structure of the DBN for the Indian Pines data set is set as 200–50 - ... - 50 - 8, which means the input layer has 200 nodes corresponding to the dimension of input data, the output layer has eight nodes corresponding to the number of classes, and all the middle layers have 50 nodes. Particulars about the number of training and testing samples are presented in **Table 1**. The performance of the DBN can be significantly improved by modifying the pretraining and fine-tuning of D-DBNs. DBN-based classification methods realizes comparatively fast inference and

(a) (b) (c)

Figure 5.
Indian Pines data set. (a) Original image produced by the mixture of three bands. (b) Ground truth with eight classes. (c) Map color.

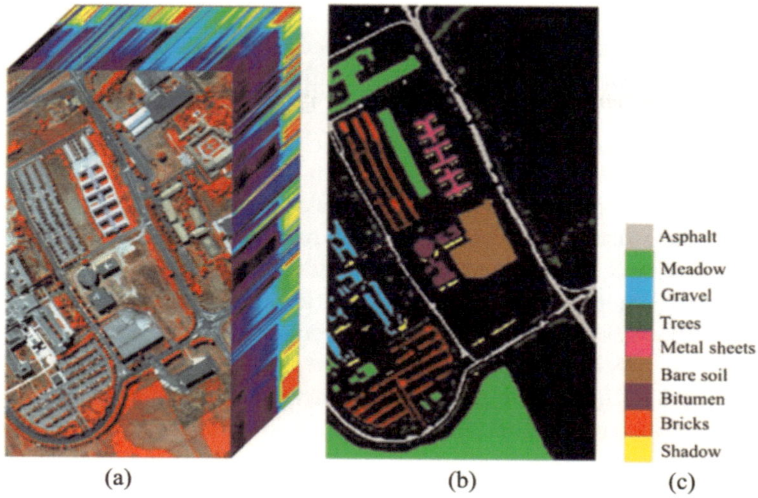

Figure 6.
University of Pavia data set. (a) Original image produced by the mixture of three bands. (b) Ground truth with nine classes. (c) Map color.

ID	Indian pines			University of Pavia		
	Class name	**Training**	**Test**	**Class name**	**Training**	**Test**
1	Corn-notill	200	1234	Asphalt	200	6431
2	Corn-mintill	200	634	Meadows	200	18,499
3	Grass-pasture	200	297	Gravel	200	1899
4	Hay-windrowed	200	289	Trees	200	2864
5	Soybean-notill	200	768	Sheets	200	1145
6	Soybean-mintill	200	2268	Bare soil	200	4829
7	Soybean-clean	200	414	Bitumen	200	1130
8	Woods	200	1094	Bricks	200	3482
9				Shadows	200	747
Total		1600			1800	40,976

Table 1.
Number of training and test samples.

competent representation of hyperspectral image and thus good classification performance.

7.2 Convolutional neural networks (CNN)

Quite a few number of neural network-based classification methods have been proposed in the literature to deal with both supervised and unsupervised nonparametric approaches [72–74]. The feedforward neural network (FN)-based classifiers are extensively used with the variation of second-order optimization-based strategies, which are faster and need fewer input parameters [75, 76]. The extreme learning machine (ELM) learning algorithm has became popular that train single hidden-layer FNs (SLFN) [77, 78]. Then, the concept has been extended to multi-hidden-layer networks [79], radial basis function (RBF) networks [80], and kernel learning [81, 82]. ELM-based networks are remarkably efficient in terms of accuracy and computational complexity and have been successfully applied as nonlinear classifiers for hyperspectral data, providing results comparable with state-of-the-art methodologies.

In recent years, convolutional neural network (CNN) has acquired auspicious achievements in remote sensing [58, 83–85]. The deep structure of CNNs allows the model to learn highly abstract feature detectors and to map the input features into representations that can clearly boost the performance of the subsequent classifiers. The advantage of such approaches over probabilistic methods result mainly from the fact that neural networks do not need prior knowledge about the statistical distribution of the classes. Their attractiveness increased because of the availability of feasible training techniques for nonlinearly separable data citepbenediktsson1990statistical, although their use has been traditionally affected by their algorithmic and training complexity [86] as well as by the number of parameters that need to be tuned.

The CNN is a multi-layer architecture with multiple stages for effective feature-extraction. Generally, each stage of CNN is composed of three layers. (i) convolutional layer (ii) nonlinearity layer and and (iii) pooling layer. The classical CNN is composed of one, two, or three feature-extraction stages, followed by one or more fully connected layers and a final classifier layer.

Convolutional layer: The input to the convolutional layer is represented as $x^i_{m,n}$, with r number of features maps x^i, each map is of size $m \times n$. The convolutional layer consists of filter banks W of size $l \times l \times q$ that connects input filter map to output filter map. The output of convolutional layer is a three-dimensional array $m_1 \times n_1 \times k$, composed of k feature maps of size $m_1 \times n_1$. The output of the convolutional layer is determined as:

$$z^s = \sum_{i=1}^{q} W_i^s * x^i + b_s \tag{32}$$

Where, b is the bias paprameter.

Nonlinearity layer: The nonlinearity layer measures the output feature map $a^s = f(z^s)$, as f(.) is usually selected to be a rectified linear unit (ReLU) $f(x) = \max(0,x)$.

Pooling Layer: The pooling layer involves executing a max operation over the activations within a small spatial region G of each feature map: $p^s_G = \max_{i \in G} a^i_s$. After the multiple feature-extraction stages, the entire network is trained with back propagation of a supervised loss function such as the classic least-squares output, and the target output γ is represented as a L-of-K vector, where K is the number of output and L is the number of layers:

$$J(\theta) = \sum_{i=1}^{N} \left(\frac{1}{2}\|h(x_i, \theta) - \gamma\|^2\right) + \lambda \sum_{l}^{L} \text{sum}\left(\|\theta^{(l)}\|^2\right), \tag{33}$$

where l indexes the layer number. Primary goal is to minimize $J(\theta)$ as a function of θ. To train the CNN, stochastic gradient descent with back propagation is exercised to optimize the function.

The three fundamental parts of a CNN are a convolutional layer, non linear function and a pooling layer. A deep CNN can be formulated by stacking several convolution layers with nonlinear operation and several pooling layers. A deep CNN can hierarchically extract the features of inputs, which tend to be invariant and robust [87]. The architecture of a deep CNN for spectral classification is shown in **Figure 7**.

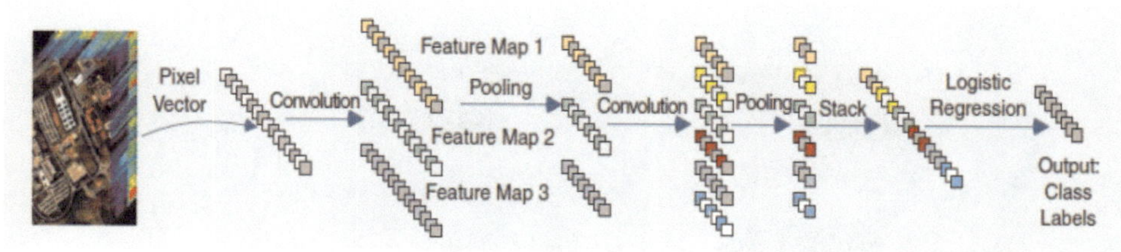

Figure 7.
A spectral classifier based on a deep CNN.

A systematic survey on deep networks for remote sensing data has been presented in [56]. In [83], CNN was investigated to exploit deep representation based on spectral signatures and the performance proved to be superior to that of SVM. The high level spatial features are extracted using CNN [88], deep CNN for pixel classification while learning unsupervised sparse features [59], deep CNN to learn pixel-pair features [89] and few more.

The performance of the HSI classification method proposed in [83] termed as deep CNN (D-CNN) is compared with a traditional SVM classifier. Two hyperspectral data sets including Indian Pines and University Of Pavia are used for the evaluation. The Indian Pines data set consists of 220 spectral channels in the 0.4–2.45 μm region of the visible and infrared spectrum with a spatial resolution of 20 m. The University of Pavia data set with a spatial coverage of 610 × 340 pixels covering the city of Pavia and has 103 spectral bands prior to water band removal. It has a spectral coverage from 0.43 to 0.86 μm and a spatial resolution of 1.3 m. All the layer parameters of these two data sets for CNN classifier are set as specified in [83]. The comparison of classification performance between D-CNN and SVM is presented in **Table 2**. **Figures 8** and **9** interpret the corresponding classification

Data set	D-CNN (%)	SVM (%)
Indian pines	90.18	87.54
University of Pavia	92.64	90.42

Table 2.
Comparison of results between the D-CNN and SVM using two data sets.

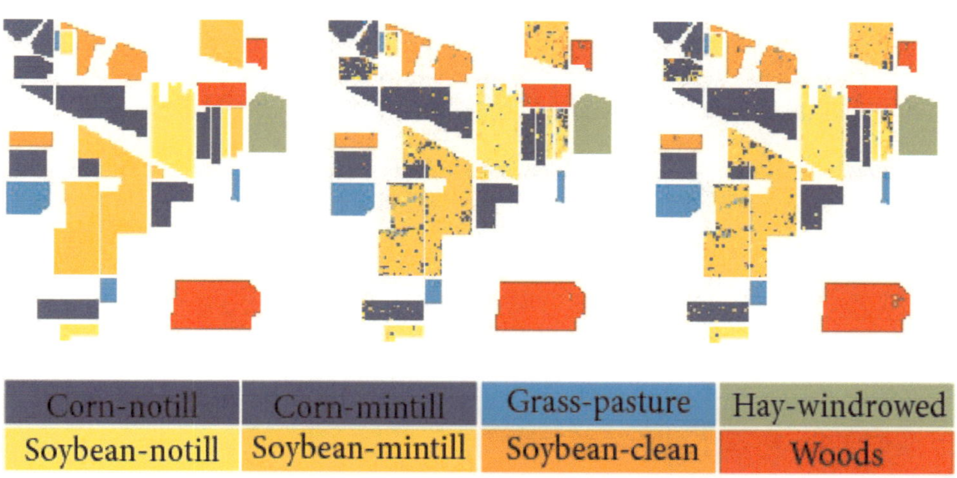

Figure 8.
RGB composition maps resulting from classification for the Indian Pines data set. From left to right: ground truth, SVM, and D-CNN.

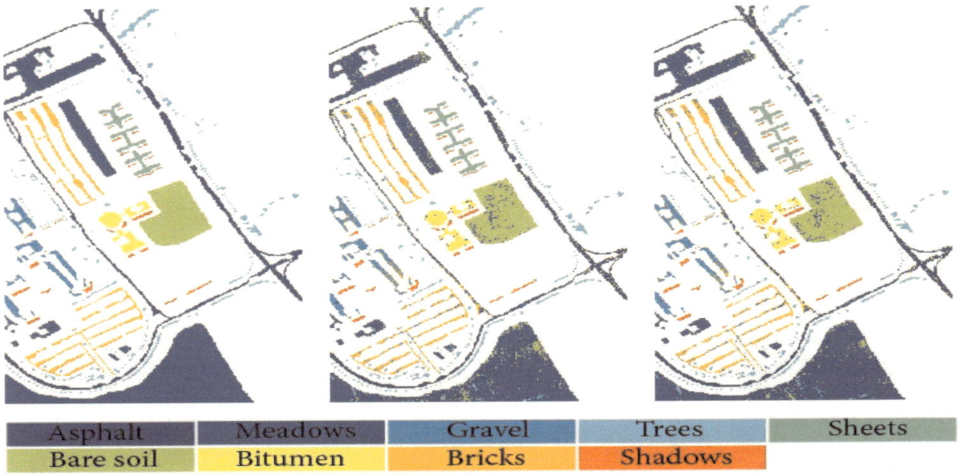

Figure 9.
Thematic maps resulting from classification for University of Pavia data set. From left to right: ground truth, SVM, and D-CNN.

maps obtained with D-CNN and SVM classifier. Furthermore, compared with traditional SVM the D-CNN classifier has higher classification accuracy for the overall data sets.

Furthermore, the application of Deep learning to hyperspectral image classification has some potential issues to be investigated.

i. Deep learning methods may lead to a serious problem called overfitting, which means that the results can be very good on the training data but poor on the test data. To deal with this issue, it is necessary to use powerful regularization methods.

ii. In contrast to natural images, the high resolution remote sensing (RS) images are complex in nature. The complexity of RS images leads to some difficulty in descriminative representation and learning features from the objects with DL.

iii. The deepaer layers in supervised networks like CNNs can learn more complex distributions. Research on appropriate depth for a DL model for a given data set is still an open research topic to be explored.

iv. Deep learning methods can be combined with other methods, such as sparse coding and ensemble learning which is another research area in hyperspectral data classification.

Author details

Rajesh Gogineni* and Ashvini Chaturvedi
National Institute of Technology Karnataka, Mangalore, India

*Address all correspondence to: rgogineni9@gmail.com

References

[1] Bioucas-Dias JM, Plaza A, Camps-Valls G, Scheunders P, Nasrabadi N, Chanussot J. Hyperspectral remote sensing data analysis and future challenges. IEEE Geoscience and remote sensing magazine. 2013;**1**(2):6-36

[2] Tong X, Xie H, Weng Q. Urban land cover classification with airborne hyperspectral data: What features to use? IEEE Journal of Selected Topics in Applied Earth Observations and Remote Sensing. 2013;**7**(10):3998-4009

[3] Gevaert CM, Suomalainen J, Tang J, Kooistra L. Generation of spectral–temporal response surfaces by combining multispectral satellite and hyperspectral uav imagery for precision agriculture applications. IEEE Journal of Selected Topics in Applied Earth Observations and Remote Sensing. 2015; **8**(6):3140-3146

[4] Yuen PW, Richardson M. An introduction to hyperspectral imaging and its application for security, surveillance and target acquisition. The Imaging Science Journal. 2010;**58**(5): 241-253

[5] Zhang L, Zhang L, Tao D, Huang X, Du B. Hyperspectral remote sensing image subpixel target detection based on supervised metric learning. IEEE Transactions on Geoscience and Remote Sensing. 2014;**52**(8):4955-4965

[6] Yang X, Yu Y. Estimating soil salinity under various moisture conditions: An experimental study. IEEE Transactions on Geoscience and Remote Sensing. 2017;**55**(5):2525-2533

[7] Shahshahani BM, Landgrebe DA. The effect of unlabeled samples in reducing the small sample size problem and mitigating the Hughes phenomenon. IEEE Transactions on Geoscience and Remote Sensing. 1994;**32**(5): 1087-1095

[8] Chi M, Bruzzone L. Semisupervised classification of hyperspectral images by svms optimized in the primal. IEEE Transactions on Geoscience and Remote Sensing. 2007;**45**(6):1870-1880

[9] Hughes G. On the mean accuracy of statistical pattern recognizers. IEEE Transactions on Information Theory. 1968;**14**(1):55-63

[10] Jin J, Wang B, Zhang L. A novel approach based on fisher discriminant null space for decomposition of mixed pixels in hyperspectral imagery. IEEE Geoscience and Remote Sensing Letters. 2010;**7**(4):699-703

[11] Zhang L, Zhang L, Tao D, Huang X. On combining multiple features for hyperspectral remote sensing image classification. IEEE Transactions on Geoscience and Remote Sensing. 2011; **50**(3):879-893

[12] Zhong Y, Zhang L. An adaptive artificial immune network for supervised classification of multi−/ hyperspectral remote sensing imagery. IEEE Transactions on Geoscience and Remote Sensing. 2012;**50**(3):894-909

[13] Melgani F, Bruzzone L. Classification of hyperspectral remote sensing images with support vector machines. IEEE Transactions on Geoscience and Remote Sensing. 2004; **42**(8):1778-1790

[14] Mohamed RM, Farag AA. Advanced algorithms for bayesian classification in high dimensional spaces with applications in hyperspectral image segmentation. In: IEEE International Conference on Image Processing. Vol. 2. IEEE; 2005. pp. II-646

[15] Camps-Valls G, Gomez-Chova L, Muñoz-Marí J, Vila-Francés J, Calpe-Maravilla J. Composite kernels for hyperspectral image classification. IEEE

Geoscience and Remote Sensing Letters. 2006;**3**(1):93-97

[16] Wold S, Esbensen K, Geladi P. Principal component analysis. Chemometrics and Intelligent Laboratory Systems. 1987;**2**(1–3):37-52

[17] Hyvärinen A, Oja E. Independent component analysis: Algorithms and applications. Neural Networks. 2000;**13**(4–5):411-430

[18] Rodarmel C, Shan J. Principal component analysis for hyperspectral image classification. Surveying and Land Information Science. 2002;**62**(2): 115-122

[19] Villa A, Benediktsson JA, Chanussot J, Jutten C. Hyperspectral image classification with independent component discriminant analysis. IEEE Transactions on Geoscience and Remote Sensing. 2011;**49**(12):4865-4876

[20] Li C, Yin J, Zhao J. Using improved Ica method for hyperspectral data classification. Arabian Journal for Science and Engineering. 2014;**39**(1):181-189

[21] Du P, Liu P, Xia J, Feng L, Liu S, Tan K, et al. Remote sensing image interpretation for urban environment analysis: Methods, system and examples. Remote Sensing. 2014;**6**(10): 9458-9474

[22] Huang X, Zhang L. A comparative study of spatial approaches for urban mapping using hyperspectral rosis images over Pavia city, northern Italy. International Journal of Remote Sensing. 2009;**30**(12):3205-3221

[23] Bajorski P. Target detection under misspecified models in hyperspectral images. IEEE Journal of Selected Topics in Applied Earth Observations and Remote Sensing. 2012;**5**(2):470-477

[24] Govender M, Chetty K, Bulcock H. A review of hyperspectral remote sensing and its application in vegetation and water resource studies. Water SA. 2007;**33**(2)

[25] Jia X, Richards JA. Efficient maximum likelihood classification for imaging spectrometer data sets. IEEE Transactions on Geoscience and Remote Sensing. 1994;**32**(2):274-281

[26] Yonezawa C. Maximum likelihood classification combined with spectral angle mapper algorithm for high resolution satellite imagery. International Journal of Remote Sensing. 2007;**28**(16):3729-3737

[27] Kuo BC, Yang JM, Sheu TW, Yang SW. Kernel-based knn and gaussian classifiers for hyperspectral image classification. In: IGARSS 2008–2008 IEEE International Geoscience and Remote Sensing Symposium. Vol. 2. IEEE; 2008. pp. II-1006

[28] Yang J-M, Yu P-T, Kuo B-C. A nonparametric feature extraction and its application to nearest neighbor classification for hyperspectral image data. IEEE Transactions on Geoscience and Remote Sensing. 2010;**48**(3):1279-1293

[29] Calin MA, Parasca SV, Manea D. Comparison of spectral angle mapper and support vector machine classification methods for mapping skin burn using hyperspectral imaging. In: Unconventional Optical Imaging. Vol. 10677. International Society for Optics and Photonics. 2018. p. 106773

[30] Bazi Y, Melgani F. Toward an optimal SVM classification system for hyperspectral remote sensing images. IEEE Transactions on Geoscience and Remote Sensing. 2006;**44**(11):3374-3385

[31] Gu Y, Feng K. Optimized laplacian SVM with distance metric learning for hyperspectral image classification. IEEE journal of selected topics in applied earth observations and remote sensing. 2013;**6**(3):1109-1117

[32] Tarabalka Y, Fauvel M, Chanussot J, Benediktsson JA. SVM-and MRF-based method for accurate classification of hyperspectral images. IEEE Geoscience and Remote Sensing Letters. 2010;7(4): 736-740

[33] Santos AB, de Albuquerque Araújo A, Menotti D. Combining multiple classification methods for hyperspectral data interpretation. IEEE Journal of Selected Topics in Applied Earth Observations and Remote Sensing. 2013;**6**(3):1450-1459

[34] Chen Y, Zhao X, Lin Z. Optimizing subspace SVM ensemble for hyperspectral imagery classification. IEEE Journal of Selected Topics in Applied Earth Observations and Remote Sensing. 2014;7(4):1295-1305

[35] Vapnik V. Statistical learning theory. New York: John Wiley & Sons Inc.; 1998

[36] Ham J, Chen Y, Crawford MM, Ghosh J. Investigation of the random forest framework for classification of hyperspectral data. IEEE Transactions on Geoscience and Remote Sensing. 2005;**43**(3):492-501

[37] Zhang Y, Cao G, Li X, Wang B. Cascaded random forest for hyperspectral image classification. IEEE Journal of Selected Topics in Applied Earth Observations and Remote Sensing. 2018;**11**(4):1082-1094

[38] Liu J, Wu Z, Wei Z, Xiao L, Sun L. Spatial-spectral kernel sparse representation for hyperspectral image classification. IEEE Journal of Selected Topics in Applied Earth Observations and Remote Sensing. 2013;**6**(6):2462-2471

[39] He L, Li J, Liu C, Li S. Recent advances on spectral–spatial hyperspectral image classification: An overview and new guidelines. IEEE Transactions on Geoscience and Remote Sensing. 2018;**56**(3):1579-1597

[40] Tang YY, Lu Y, Yuan H. Hyperspectral image classification based on three-dimensional scattering wavelet transform. IEEE Transactions on Geoscience and Remote Sensing. 2015; **53**(5):2467-2480

[41] Rajadell O, García-Sevilla P, Pla F. Spectral–spatial pixel characterization using gabor filters for hyperspectral image classification. IEEE Geoscience and Remote Sensing Letters. 2013;**10**(4): 860-864

[42] Bourennane S, Fossati C, Cailly A. Improvement of classification for hyperspectral images based on tensor modeling. IEEE Geoscience and Remote Sensing Letters. 2010;7(4):801-805

[43] Fauvel M, Tarabalka Y, Benediktsson JA, Chanussot J, Tilton JC. Advances in spectral-spatial classification of hyperspectral images. Proceedings of the IEEE. 2013;**101**(3): 652-675

[44] Aptoula E, Lefèvre S. A comparative study on multivariate mathematical morphology. Pattern Recognition. 2007;**40**(11):2914-2929

[45] Li J, Bioucas-Dias JM, Plaza A. Spectral–spatial classification of hyperspectral data using loopy belief propagation and active learning. IEEE Transactions on Geoscience and Remote Sensing. 2013;**51**(2):844-856

[46] Rath G, Sahoo A. A comparative study of some greedy pursuit algorithms for sparse approximation. In: 2009 17th European Signal Processing Conference; 2009. pp. 398-402

[47] Ni D, Ma H. Hyperspectral image classification via sparse code histogram. IEEE Geoscience and Remote Sensing Letters. 2015;**12**(9):1843-1847

[48] Fang L, Li S, Kang X, Benediktsson JA. Spectral–spatial classification of hyperspectral images

with a superpixel-based discriminative sparse model. IEEE Transactions on Geoscience and Remote Sensing. 2015; **53**(8):4186-4201

[49] Chen Y, Nasrabadi NM, Tran TD. Hyperspectral image classification using dictionary-based sparse representation. IEEE Transactions on Geoscience and Remote Sensing. 2011;**49**(10):3973-3985

[50] Zhang H, Li J, Huang Y, Zhang L. A nonlocal weighted joint sparse representation classification method for hyperspectral imagery. IEEE Journal of Selected Topics in Applied Earth Observations and Remote Sensing. 2013; 7(6):2056-2065

[51] Wang Z, Nasrabadi NM, Huang TS. Spatial–spectral classification of hyperspectral images using discriminative dictionary designed by learning vector quantization. IEEE Transactions on Geoscience and Remote Sensing. 2013;**52**(8):4808-4822

[52] Soltani-Farani A, Rabiee HR, Hosseini SA. Spatial-aware dictionary learning for hyperspectral image classification. IEEE Transactions on Geoscience and Remote Sensing. 2014; **53**(1):527-541

[53] Sun X, Nasrabadi NM, Tran TD. Task-driven dictionary learning for hyperspectral image classification with structured sparsity constraints. IEEE Transactions on Geoscience and Remote Sensing. 2015;**53**(8):4457-4471

[54] Landgrebe DA. Signal Theory Methods in Multispectral Remote Sensing. Vol. 29. John Wiley & Sons; 2005

[55] Ghamisi P, Plaza J, Chen Y, Li J, Plaza AJ. Advanced spectral classifiers for hyperspectral images: A review. IEEE Geoscience and Remote Sensing Magazine. 2017;**5**(1):8-32

[56] Zhang L, Zhang L, Du B. Deep learning for remote sensing data: A technical tutorial on the state of the art. IEEE Geoscience and Remote Sensing Magazine. 2016;**4**(2):22-40

[57] Chen Y, Zhao X, Jia X. Spectral–spatial classification of hyperspectral data based on deep belief network. IEEE Journal of Selected Topics in Applied Earth Observations and Remote Sensing. 2015;**8**(6):2381-2392

[58] Chen Y, Lin Z, Zhao X, Wang G, Gu Y. Deep learning-based classification of hyperspectral data. IEEE Journal of Selected topics in applied earth observations and remote sensing. 2014; **7**(6):2094-2107

[59] Romero A, Gatta C, Camps-Valls G. Unsupervised deep feature extraction for remote sensing image classification. IEEE Transactions on Geoscience and Remote Sensing. 2016;**54**(3):1349-1362

[60] Hinton GE, Osindero S, Teh Y-W. A fast learning algorithm for deep belief nets. Neural Computation. 2006;**18**(7): 1527-1554

[61] Freund Y, Haussler D. Unsupervised learning of distributions on binary vectors using two layer networks. In: Advances in Neural Information Processing Systems. 1992. pp. 912-919

[62] Liu P, Zhang H, Eom KB. Active deep learning for classification of hyperspectral images. IEEE Journal of Selected Topics in Applied Earth Observations and Remote Sensing. 2016;**10**(2):712-724

[63] Hinton GE, Srivastava N, Krizhevsky A, Sutskever I, Salakhutdinov RR. Improving neural networks by preventing co-adaptation of feature detectors. arXiv preprint arXiv:1207.0580. 2012

[64] Shaham U, Cheng X, Dror O, Jaffe A, Nadler B, Chang J, et al. A deep learning approach to unsupervised ensemble learning. In: International

Conference on Machine Learning; 2016. pp. 30-39

[65] Xiong H, Rodríguez-Sánchez AJ, Szedmak S, Piater J. Diversity priors for learning early visual features. Frontiers in Computational Neuroscience. 2015;**9**: 104

[66] Qiao M, Bian W, Da Xu RY, Tao D. Diversified hidden Markov models for sequential labeling. IEEE Transactions on Knowledge and Data Engineering. 2015;**27**(11):2947-2960

[67] Zhong P, Peng N, Wang R. Learning to diversify patch-based priors for remote sensing image restoration. IEEE Journal of Selected Topics in Applied Earth Observations and Remote Sensing. 2015;**8**(11):5225-5245

[68] Kwok JT, Adams RP. Priors for diversity in generative latent variable models. In: Advances in Neural Information Processing Systems. 2012. pp. 2996-3004

[69] Xie P, Deng Y, Xing E. Diversifying restricted boltzmann machine for document modeling. In: Proceedings of the 21th ACM SIGKDD International Conference on Knowledge Discovery and Data Mining. ACM. 2015. pp. 1315-1324

[70] Zhong P, Gong Z, Li S, Schönlieb CB. Learning to diversify deep belief networks for hyperspectral image classification. IEEE Transactions on Geoscience and Remote Sensing. 2017;**55**(6):3516-3530

[71] Gong B, Chao W-L, Grauman K, Sha F. Diverse sequential subset selection for supervised video summarization. In: Advances in Neural Information Processing Systems. papers.nips.cc. 2014. pp. 2069-2077

[72] Merényi E, Farrand WH, Taranik JV, Minor TB. Classification of hyperspectral imagery with neural networks: Comparison to conventional tools. EURASIP Journal on Advances in Signal Processing. 2014;**2014**(1):71

[73] Del Frate F, Pacifici F, Schiavon G, Solimini C. Use of neural networks for automatic classification from high-resolution images. IEEE Transactions on Geoscience and Remote Sensing. 2007; **45**(4):800-809

[74] Ratle F, Camps-Valls G, Weston J. Semisupervised neural networks for efficient hyperspectral image classification. IEEE Transactions on Geoscience and Remote Sensing. 2010; **48**(5):2271-2282

[75] Hagan MT, Menhaj MB. Training feedforward networks with the marquardt algorithm. IEEE Transactions on Neural Networks. 1994; **5**(6):989-993

[76] Rumelhart DE, Hinton GE, Williams RJ, et al. Learning representations by back-propagating errors. Cognitive modeling. 1988;**5**(3):1

[77] Huang G-B, Zhu Q-Y, Siew C-K. Extreme learning machine: Theory and applications. Neurocomputing. 2006;**70** (1–3):489-501

[78] Huang G, Huang G-B, Song S, You K. Trends in extreme learning machines: A review. Neural Networks. 2015;**61**:32-48

[79] Tang J, Deng C, Huang G-B. Extreme learning machine for multilayer perceptron. IEEE transactions on neural networks and learning systems. 2015;**27**(4):809-821

[80] Huang GB, Siew CK. Extreme learning machine: Rbf network case. In: ICARCV 2004 8th Control, Automation, Robotics and Vision Conference; 2004. Vol. 2. IEEE; 2004. pp. 1029-1036

[81] Huang G-B. An insight into extreme learning machines: Random neurons,

random features and kernels. Cognitive Computation. 2014;**6**(3):376-390

[82] Zhou Y, Peng J, Chen CP. Extreme learning machine with composite kernels for hyperspectral image classification. IEEE Journal of Selected Topics in Applied Earth Observations and Remote Sensing. 2014;**8**(6): 2351-2360

[83] Hu W, Huang Y, Wei L, Zhang F, Li H. Deep convolutional neural networks for hyperspectral image classification. Journal of Sensors. 2015; **2015**

[84] Makantasis K, Karantzalos K, Doulamis A, Doulamis N. Deep supervised learning for hyperspectral data classification through convolutional neural networks. In: 2015 IEEE International Geoscience and Remote Sensing Symposium (IGARSS). IEEE; 2015. pp. 4959-4962

[85] Chen Y, Jiang H, Li C, Jia X, Ghamisi P. Deep feature extraction and classification of hyperspectral images based on convolutional neural networks. IEEE Transactions on Geoscience and Remote Sensing. 2016;**54**(10):6232-6251

[86] Richards JA. Analysis of remotely sensed data: The formative decades and the future. IEEE Transactions on Geoscience and Remote Sensing. 2005; **43**(3):422-432

[87] Bengio Y, Courville A, Vincent P. Representation learning: A review and new perspectives. IEEE Transactions on Pattern Analysis and Machine Intelligence. 2013;**35**(8):1798-1828

[88] Zhao W, Du S. Spectral–spatial feature extraction for hyperspectral image classification: A dimension reduction and deep learning approach. IEEE Transactions on Geoscience and Remote Sensing. 2016;**54**(8):4544-4554

[89] Li W, Wu G, Zhang F, Du Q. Hyperspectral image classification using deep pixel-pair features. IEEE Transactions on Geoscience and Remote Sensing. 2017;**55**(2):844-853

Chapter 3

Hyperspectral Image Super-Resolution Using Optimization and DCNN-Based Methods

Xian-Hua Han

Abstract

Reconstructing a high-resolution (HR) hyperspectral (HS) image from the observed low-resolution (LR) hyperspectral image or a high-resolution multispectral (RGB) image obtained using the exiting imaging cameras is an important research topic for capturing comprehensive scene information in both spatial and spectral domains. The HR-HS hyperspectral image reconstruction mainly consists of two research strategies: optimization-based and the deep convolutional neural network-based learning methods. The optimization-based approaches estimate HR-HS image via minimizing the reconstruction errors of the available low-resolution hyperspectral and high-resolution multispectral images with different constrained prior knowledge such as representation sparsity, spectral physical properties, spatial smoothness, and so on. Recently, deep convolutional neural network (DCNN) has been applied to resolution enhancement of natural images and is proven to achieve promising performance. This chapter provides a comprehensive description of not only the conventional optimization-based methods but also the recently investigated DCNN-based learning methods for HS image super-resolution, which mainly include spectral reconstruction CNN and spatial and spectral fusion CNN. Experiment results on benchmark datasets have been shown for validating effectiveness of HS image super-resolution in both quantitative values and visual effect.

Keywords: hyperspectral imaging, image super-resolution, optimization-based approach, deep convolutional neural network (DCNN), spectral reconstruction, spatial and spectral fusion

1. Introduction

Hyperspectral (HS) imaging simultaneously obtains a set of images of the same scene on a large number of narrow-band wavelengths which can effectively describe the spectral distribution for every scene point and provide intrinsic and discriminative spectral information of the scene. The acquired dense spectral bands of data are capable to benefit for numerous applications, including object recognition and segmentation [1–9], medical image analysis [10], and remote sensing [11–15], to name a few. Although with the availability of the abundant spectral information with HS imaging, it generally results in much low spatial resolution compared with ordinary panchromatic and RGB images since photon collection in HS sensors is performed in a much larger spatial region for guaranteeing sufficiently high signal-to-noise ratio. The low spatial resolution in the HS images leads to high spectral mixing of different materials in a scene and greatly affects the performance of scene analysis and understanding. Therefore, the reconstruction of high-resolution hyperspectral (HR-HS) image using image processing and machine leaning techniques has attracted a lot of attention.

Especially in remote sensing field, a low-resolution (LR) multispectral or HS image is usually available accompanying with a HR single-channel panchromatic image, and the fusion of these two images is generally known as the pan-sharpening technique [16–39]. Motivated by the fact that human vision is more sensitive to luminance, traditional pan-sharpening technique mainly concentrated the reliable illumination restoration via substituting the calculated component of the LR-HS image with the HR information of panchromatic image via sue saturation exploring and principle component analysis. However, these simple approaches avoidably cause spectral distortion in the resulting image. Recently, the HS image super-resolution actively investigates the optimization methods for minimizing the reconstruction error of the available LR-HS and HR-MS (HR-RGB) images [16–30], which manifested impressive performance. The basic idea of these optimization-based approaches assumes that the spectrum can be represented as matrix decomposition with different constraints such as representation sparsity, spectral physical properties, spatial context similarity, and composited matrixes, which are iteratively optimized for more accurate approximating the observed images. Recently, the matrix factorization and spectral unmixing [40–43]-based HS image super-resolution, which are mainly motivated by the fact the HS observations can be represented by a linear combination of the reflectance function basis (the spectral signatures of the pure materials) and the weight vector denoting the fractions of the pure materials on the spectral response is assumed sparse, have been actively investigated [16, 17, 27, 28]. A coupled nonnegative matrix factorization (CNMF) by Yokoya et al. [19], inspired by the physical property of nonnegative weights for the linear combination, has been proposed to estimate the HR-HS image from a pair of HR-MS and LR-HS images. Although the CNMF approach provided acceptable spectral recovery performance, its solution is usually not unique [44], which cannot always lead to unsatisfied spectral recovery results. Lanaras et al. [10] proposed to integrate coupled spectral unmixing strategy into HS super-resolution and conducted optimization procedure with the proximal alternating linearized minimization method, which requires the good initial points of the two decomposed reflectance signatures and the fraction vectors for providing impressive results. Furthermore, taking consideration of the physical meaning of the spectral linear combination on the reflectance signatures and the implementation effectiveness, most work generally assumes that the number of the pure materials in the observed scene is smaller than the spectral band number, which is not always satisfied in the real application.

Motivated by the successful applications of the sparse representation on the natural image analysis [14, 15] such as image de-noising, super-resolution, and representation, the sparsity-promoting approaches without considering explicitly the physical meaning constraint on the reflection signature (basis) and thus permitting over-complete basis have widely been applied for HS super-resolution [18, 19]. Inspired by the work in the general RGB image analysis with sparse representation, Grohnfeldt et al. [11] explored a joint sparse representation for HS image super-resolution. Via learning the corresponding HS and MS (RGB) patch dictionaries using the prepared pairs, this work assumed the same sparse coefficients of the corresponding MS and HS patch dictionary, and thus, these can be calculated with only the MS input patch. However the above procedure was conducted on each individual band, which mainly considered the well reconstruction of the local structure (patch) and completely ignored the spectral correlation between channels. Therefore, several other works [19, 22] investigated the sparse spectral representation via conducting reconstruction of all band spectra instead of the local structure on each individual band. Akhtar et al. [13] explored a sparse spatiospectral representation via calculating the optimized sparse coefficients of each spectral pixel but assuming the same used atoms for the pixels in a local grid region to integrate the spatial structure. For calculation effectiveness, a generalized simultaneous orthogonal matching pursuit (G-SOMP) was proposed for estimating the sparse coefficients in [22]. Later, the same research group integrated the sparse representation and the Bayesian dictionary learning algorithm for improving the HS image super-resolution performance and manifested its effectiveness. Dong et al. [21] proposed a nonnegative structured sparse representation (NSSR) approach for taking consideration of the spatial structure and then conducted optimization procedure with the alternative direction multiplier method (ADMM) technique. NSSR achieved a large margin on HS image recovery performance compared with the other state-of-the-art approaches. Furthermore, Han et al. [45] proposed to recover the HR-HS output via minimizing the coupled reconstruction error of the available LR-HR and HR-RGB images with the following constraints, (1) the sparse representation with over-complete spectral dictionary in the coupled unmixing strategy [17] and (2) the self-similarity of the sparse spectral representation in the global structures and the local spectra existed in the available HR-RGB image, which further improved the HS image recovery performance in both visual and quality aspects.

Deep convolutional neural networks (CNNs) have recently shown great success in various image processing and computer vision applications. CNN has also been applied to RGB image super-resolution and achieved promising performance. Dong et al. [46] proposed a three-layer CNN architecture (SRCNN), which demonstrates about 0.5–1.5 db improvement and much lower computational cost compared with the popularly used sparse-based methods, and they further extended SRCNN to be capable of directly dealing with the available LR images without mathematical upsampling operation, called as fast SRCNN. Kim et al. [47] exploited a very deep CNN architecture based on VGG-net architecture and concentrated on only estimating the missing high-frequency image (residual image). Ledig et al. integrated two different types of networks, generate network and discriminate network (called as GAN), for estimating much sharper HR image. For applying CNN to HSI SR, Li et al. [48] applied similar structures of SRCNN to super-resolve HSI only from the LR-HS image. These CNN architectures take only the LR image as input, and the expanding factor of resolution enhancement is theoretically limited to be lower than 8 in both height and width. There are also several works exploring CNN-based method with variant backbone architectures to expand the spectral resolution with only HR-RGB image as input [49, 50]. This chapter introduces several research works based on DCNN learning for HS image reconstruction.

On the other hand, regarding to the use of the observed data, the HR-HS image reconstruction can be divided into three research directions: (1) spatial resolution enhancement from hyperspectral imaging, (2) spectral resolution enhancement from RGB imaging, and (3) fusion method based on the observed HR-RGB and low-resolution (LR) HS images of the same scene. Spatial resolution enhancement has popularly been used on single natural image super-resolution [46, 47], and impressive performance has been achieved especially with the deep learning method in the resolution expanding factor from 2 to 4. The deep convolutional neural network (DCNN) has also been adopted for predicting the HR-HS image from a single LR-HS image [48] and validated feasibility of HS image super-resolution for small

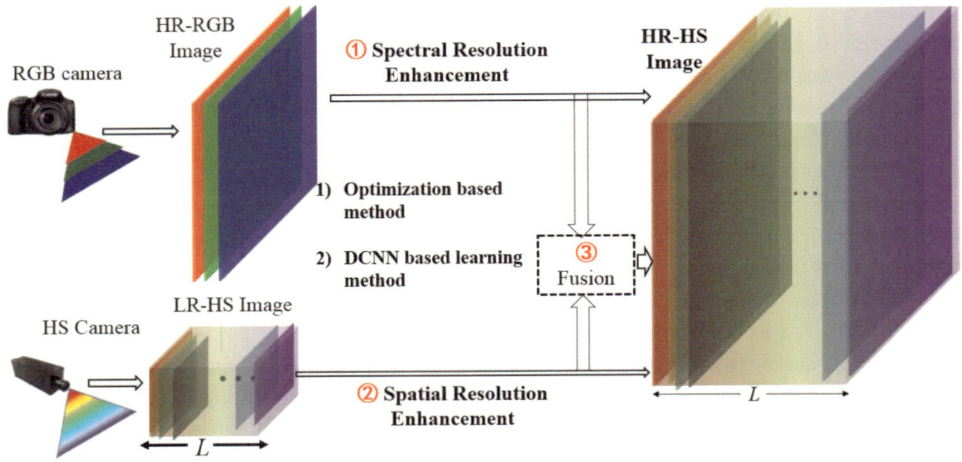

Figure 1.
The basic concept of the hyperspectral image super-resolution.

expanding factor. However, the spatial resolution of the available HS image is considerably low compared with the commonly observed RGB image, and then the expanding factor for HR-HS image reconstruction is required to be large enough, for example, more than 10 in horizontal and vertical directions, respectively. Thus, the reconstructed HS image with acceptable quality usually cannot reach the required spatial resolution for different applications. The spectral resolution enhancement for RGB-to-spectrum reconstruction [49, 50] has recently become a hot research line with a single RGB image, which can be lightly collected with a low-price visual sensor. Although the impressive potential of the RGB-spectrum reconstruction is evaluated, there has still large space for performance improving in real applications. Fusing a LR-HS image with the corresponding HR-RGB image to obtain a HR-HS image has shown promising performance [18, 19, 22, 30] compared to spatial and spectral resolution enhancement methods. It is usually solved as an optimization problem with prior knowledge such as sparsity representation and spectral physical properties as constraints, which needs comprehensive analysis of the target scene previously and would be varied scene by scene. Motivated by the amazing performance of the DCNN in natural image super-resolution, Han etc. [51] proposed a spatial and spectral fusion network (SSF-Net) for the HR-HS image reconstruction and validated the better results of the SSF-Net in spite of the simple concatenation of the upsampled LR-HS image and the HR-RGB image. However, the upsampling of the LR-HS image and the simple concatenation cannot effectively integrate the existed spatial structure and spectral property but would lead to computational cost. In addition, precise alignment is needed for the input of LR-HS and HR-RGB images and is extremely difficult due to the large difference of spatial resolution in the LR-HS and HR-RGB images. This chapter introduces several advanced DCNN-based learning methods for hyperspectral image super-resolution and manifests the impressive performance for benchmark datasets. The basic concept of the hyperspectral image super-resolution is shown in **Figure 1**.

2. Problem formulation of HS image super-resolution

The goal of HS image super-resolution is to recover a HR-HS image $\mathbf{Z}' \in \mathbb{R}^{W \times H \times L}$, where L denotes the spectral band number and W and H denote the image width and height, respectively, from a HR-MS image $\mathbf{Y}' \in \mathbb{R}^{W \times H \times l}$ $(l \ll L)$

and a LR-HS image $\mathbf{X}' \in \mathbb{R}^{w \times h \times L}$ $(w \ll W, h \ll H)$. The common used HR-MS image in the HS image SR scenario is generally a RGB image with $l = 3$ spectral bands. The matrix forms of $\mathbf{Z}'$, $\mathbf{X}'$, and $\mathbf{Y}'$ are denoted as $\mathbf{Z} \in \mathbb{R}^{L \times N}$ $(N = W \times H)$, $\mathbf{X} \in \mathbb{R}^{L \times M}$ $(M = w \times h)$, and $\mathbf{Y} \in \mathbb{R}^{3 \times N}$, respectively. Both $\mathbf{X}$ (LR-HS) and $\mathbf{Y}$ (HR-RGB) can be expressed as a linear transformation from $\mathbf{Z}$ (the desired HS image) as:

$$\mathbf{X} = \mathbf{ZD}, \mathbf{Y} = \mathbf{RZ} \tag{1}$$

where $\mathbf{D} \in \mathbb{R}^{N \times M}$ is the decimation matrix, which blurs and down-samples the HR-HS image to form the LR-HS image, and $\mathbf{R} \in \mathbb{R}^{3 \times L}$ represents the RGB camera spectral response functions that maps the HR-HS image to the HR-RGB image. With the given $\mathbf{X}$ and $\mathbf{Y}$, $\mathbf{Z}$ can be estimated by minimizing the following reconstruction error:

$$\hat{\mathbf{Z}} = \mathbf{arrmin} \|\mathbf{X} - \mathbf{ZD}\|_F^2 + \|\mathbf{Y} - \mathbf{RZ}\|_F^2 \tag{2}$$

where $\|\cdot\|_F$ denotes the Frobenius norm. Via minimizing the reconstruction errors of the observed LR-HSI, $\mathbf{X}$, and the HR-RGB image, $\mathbf{Y}$, in Eq. (2), we attempt to recover the HR-HSI, $\mathbf{Z}$. The intuitive way to solve Eq. (2) is to adopt an optimization-based strategy to minimize Eq. (2) for providing an estimation of the HR-HSI, $\mathbf{Z}$. This chapter firstly explores the alternative back-projection (ABP) algorithm to iteratively update the HR-HSI, $\mathbf{Z}$, aiming at minimizing Eq. (2). Back-projection [12] is well-known as the efficient iterative procedure to minimize the reconstruction error. Since the back-projection requires an initial estimation for updating the next $\boldsymbol{Z_t}$, we simply upsample the LR-HS image $\mathbf{X}$ as the initial state, $\boldsymbol{Z_0} = Up(\mathbf{X})$. The alternative update for $\boldsymbol{Z_t}$ at the *t-th* step is formulated as:

$$\begin{aligned} Z'_{\mathbf{t}} &= Z_{t-1} + \lambda_1 \mathbf{R}^{-1} * (\mathbf{Y} - \mathbf{R}Z_{\mathbf{t}-1}) \\ Z_{\mathbf{t}} &= Z'_t + \lambda_2 D * \left(\mathbf{D}^{\mathbf{T}} * D\right)^{-1} \left(X - Z'_t D\right) \end{aligned} \tag{3}$$

where $(\cdot)^T$ denotes the transpose operation of a matrix and $(\cdot)^{-1}$ represents the inverse operation of a matrix. λ_1 and λ_2 denote the hyper-parameters for controlling the updating weights. After the predefined number of alternative iterations, it is prospected to obtain an estimated HR-HSI. $\mathbf{Z}$, for well reconstructing the observed LR-HSI, $\mathbf{X}$, and HR-RGB image, $\mathbf{Y}$.

Since the number of the unknowns ($N*L$) is much larger than the number of available measurements ($M*L + 3*N$), the above optimization problem is highly ill-posed, and proper regularization terms are required to narrow the solution space and ensure stable estimation. A widely adopted constraint is that each pixel spectral $\mathbf{z}_n \in \mathbb{R}^L$ of $\mathbf{Z}$ lies in a low-dimensional space, and it can be decomposed as [30]:

$$\mathbf{z}_n = \sum_{k=1}^{K} \mathbf{b}_\mathbf{k} \alpha_{k,n} \textit{ subject to: } b_{i,k} \geq 0, \alpha_{k,n} \geq 0, \sum_{k=1}^{K} \alpha_{k,n} = 1 \tag{4}$$

where $\mathbf{B} \in \mathbb{R}^{L \times K} = [\mathbf{b}_1, \mathbf{b}_2, \cdots, \mathbf{b}_\mathbf{K}]$ is the set of all spectral signatures ($\mathbf{b}_\mathbf{k}$, also called as the *k-th* endmember) of K distinct materials. $\boldsymbol{\alpha_n}$ represents the fractional abundance of all K materials for the n-th pixel. Taking consideration of the physical property on the spectral reflectance, the elements in the spectral signatures and the fractional abundance are nonnegative as shown in the first and second constraint terms of Eq. (4), and the summation of abundance vector for each pixel is one.

According to $\mathbf{Y} = \mathbf{RZ}$, each pixel $\mathbf{y}_n \in \mathbb{R}^3$ in the HR-RGB image can be decomposed as:

$$\mathbf{y}_n = \mathbf{Rz}_n = \mathbf{RB}\boldsymbol{\alpha}_n = \hat{\mathbf{B}}\boldsymbol{\alpha}_n \tag{5}$$

where $\hat{\mathbf{B}}$ denotes the RGB spectral dictionary obtained via transforming the HS dictionary $\mathbf{B}$ with camera spectral function $\mathbf{R}$. With a corresponding set of the previously learned spectral dictionaries, $\hat{\mathbf{B}}$ and $\mathbf{B}$, the sparse fractional vector $\boldsymbol{\alpha}_n$ is able to be estimated from the HR-RGB pixel $\mathbf{y}_n$ only.

The matrix representation forms of Eqs. (4) and (5) can be formulated as:

$$\mathbf{Z} = \mathbf{BA}, \mathbf{Y} = \hat{\mathbf{B}}\mathbf{A} \tag{6}$$

where $\mathbf{A} = [\boldsymbol{\alpha}_1, \boldsymbol{\alpha}_2, \cdots, \boldsymbol{\alpha}_N] \in \mathbb{R}_+^{K\times N}$ is a nonnegative sparse coefficient matrix. Substituting Eq. (4) into Eq. (2), we obtain the nonnegative constrains on both $\mathbf{B}$ and $\hat{\mathbf{B}}\mathbf{A}$, which are applied in the same manner as in Eq. (2). Unless otherwise noted, the nonnegative constraint is imposed on both dictionary and sparse matrix in the following deductions:

$$\{\mathbf{B}^*, \mathbf{A}^*\} = \mathbf{argmin}_{\mathbf{B},\mathbf{A}} \|\mathbf{X} - \mathbf{BAD}\|_F^2 + \|\mathbf{Y} - \hat{\mathbf{B}}\mathbf{A}\|_F^2 \tag{7}$$

The goal of Eq. (7) is to solve both spectral dictionary $\mathbf{B}$ and coefficient matrix $\mathbf{A}$ with proper regularization terms to achieve stable and accurate solution.

3. Self-similarity constrained sparse representation for HS image super-resolution

The complete pipeline of self-constrained sparse representation for HS image super-resolution is illustrated in **Figure 2**. The main contribution of this method is to propose a nonnegative sparse representation coupled with self-similarity

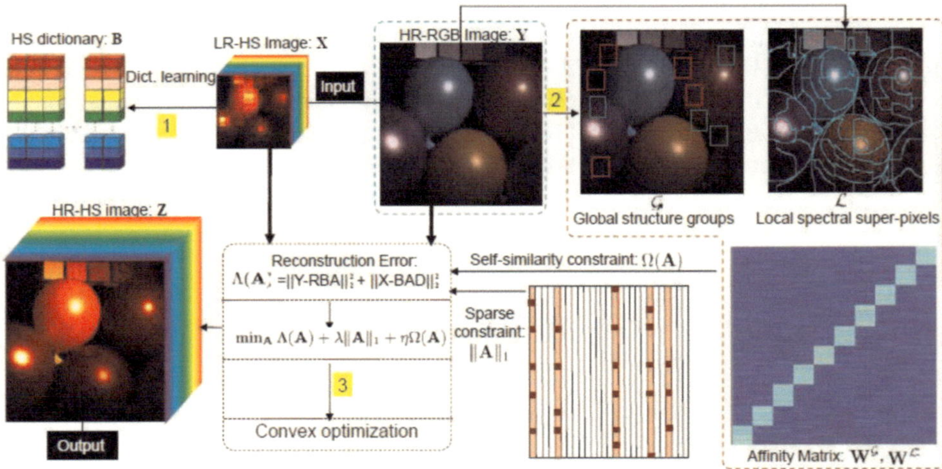

Figure 2.
Schematics of self-similarity constrained sparse representation for HS image super-resolution: (1) learn the HS dictionary B from the input LR-HS image X, (2) explore self-similarity of the global-structure and local-spectral, (3) convex optimization of the objective function with sparse and self-similarity constrains on the sparse matrix A for estimating the required HR-HS image.

constraint to regularize the solution of Eq. (7). Denoting $\mathbf{\Lambda}(\mathbf{B}, \mathbf{A}) = \|\mathbf{X} - \mathbf{BAD}\|_F^2 + \|\mathbf{Y} - \hat{\mathbf{B}}\mathbf{A}\|_F^2$, two additional terms are added to Eq. (7) as:

$$\{\mathbf{B}^*, \mathbf{A}^*\} = \mathbf{arrmin}_{\mathbf{B},\mathbf{A}}\mathbf{\Lambda}(\mathbf{B}, \mathbf{A}) + \lambda\|\mathbf{A}\|_1 + \eta\mathbf{\Omega}(\mathbf{A}) \tag{8}$$

where $\|\mathbf{A}\|_1$ denotes the sparse constrained term on the coefficient matrix and $\mathbf{\Omega}(\mathbf{A})$ represents the self-similarity regularized term. λ and η are the hyper-parameters, for controlling the contribution of the two constrained terms. Our study solves Eq. (8) with the following three steps: (1) online learning the HS dictionary from the input LR-HS image, (2) exploring the self-similarity properties of the global-structure and local-spectral self-similarity from the input HR-RGB image, and (3) conducting the convex optimization with the previously learned HS dictionary and the extracted self-similarity for estimating the HR-HS image. Next, we will describe the details of the above procedures in the following three subsections.

3.1 Online HS dictionary learning

Since different materials would have very large variety of the HS reflectance, learning a common HS dictionary for various scenes with different materials would lead to considerable spectral distortion. In order to obtain a set of adaptive HS dictionary for well reconstructing the pixel spectra, this study conducts the learning procedure directly using the observed LR-HS image $\mathbf{X}$ in an online manner. The objective function to build the HS dictionary for representing the pixel spectra is formulated as follows:

$$\left\{\mathbf{B}^*, \hat{\mathbf{A}}^*\right\} = \mathbf{argmin}_{\mathbf{B},\mathbf{A}}\left\|\mathbf{X} - \mathbf{B}\hat{\mathbf{A}}\right\|_F^2 + \lambda\left\|\hat{\mathbf{A}}\right\|_1 \tag{9}$$

where $\hat{\mathbf{A}}$ is the sparse matrix for the pixels in the LR-HS image. In our study, we also impose the nonnegative constraints on both sparse matrix $\hat{\mathbf{A}}$ and spectral dictionary $\mathbf{B}$, and thus, the existing dictionary learning method such as K-SVD cannot be applied for our optimization problem. We follow the optimization algorithm [21] and adopt ADMM technique to transform the constrained dictionary learning problem into an unconstrained version. The unconstrained dictionary learning problem is then solved with alternative optimization algorithm. After obtaining the HS dictionary $\mathbf{B}^*$ via optimizing Eq. (9) with the observed LR-HS image, we would only optimize $\mathbf{A}$ to solve Eq. (8) via fixing $\mathbf{B}^*$.

3.2 Extraction of self-similarity constraint

The regularization term $\mathbf{\Omega}(\mathbf{A})$ in Eq. (8) is formulated with two types of self-similarities, which are extracted from the HR-RGB image (see **Figure 2** for illustration):

- **Global-structure self-similarity**: Since pixels with similar spatial structure, which are represented as the concatenated RGB spectra within a local square windows, share similar hyperspectral information, thus the sparse vectors for reconstructing the hyper-spectra of these pixels would also be similar; this applies for both nearby patches and nonlocal patches in the whole image plane, and we name these as global-structure self-similarity.

- **Local-spectral self-similarity**: Since pixels in a local region have the same material with RGB values in the HR-RGB image, the sparse vector for different HR pixels is similar in a local region (superpixel). Note the superpixel is usually not a square patch.

The global-structure self-similarity is represented by global-structure groups $\mathcal{G} = \{g_1, g_2, \cdots, g_P\}$ (in total P groups), which are obtained by clustering all similar patches (spatial structure) in the HR-RGB image with K-means; $\boldsymbol{g_p}$ (each $\boldsymbol{g_p}$ may have different length) is a vector consisting of the pixel indices in the p-th group. The local-spectral self-similarity is formulated as the superpixels $\mathcal{L} = \{\boldsymbol{l}_1, \boldsymbol{l}_2, \quad, \boldsymbol{l}_Q\}$ (in total Q superpixels) obtained via SLIC superpixel segmentation method; $\boldsymbol{l_q}$ is also a vector composed with the pixel indices in the q-th superpixel. Since the pixels in the same global-structure group have similar spectral-spatial structure, we calculate the sparse vector of any pixel in a given group by a weighted average of the sparse matrix for all pixels in this group. Similarly, the sparse vector of a pixel can also be approximated by a weighted average of the sparse matrix for all pixels in the same local-spectral superpixel. With both self-similarity constraints, the sparse vector for the n-th pixel can be formulated as:

$$\boldsymbol{\alpha_n} = \gamma \sum_{i \in g_p} \mathrm{w}^{\mathcal{G}}_{\mathrm{n,i}} \boldsymbol{\alpha_i} + (1 - \gamma) \sum_{j \in l_q} \mathrm{w}^{\mathcal{L}}_{\mathrm{n,j}} \boldsymbol{\alpha_j} \quad \text{with } \mathrm{n} \in \mathrm{g_p} \wedge \mathrm{n} \in \mathrm{l_q} \tag{10}$$

where $\mathrm{w}^{\mathcal{G}}_{\mathrm{n,i}}$ is the global-structure weight for the n-th sparse vector $\boldsymbol{\alpha_n}$; it adjusts and merges the contribution of the i-th sparse vector $\boldsymbol{\alpha_i}$ belonging to the same global-structure group. Analoguely, $\mathrm{w}^{\mathcal{L}}_{\mathrm{n,j}}$ weights the j-th sparse vector $\boldsymbol{\alpha_j}$ belonging to the same local-spectral superpixel. And γ is a parameter for balancing the contribution between the global-structure and local-spectral self-similarity.

To be more specific, $\mathrm{w}^{\mathcal{G}}_{\mathrm{n,i}}$ ($0 < \mathrm{w}^{\mathcal{G}}_{\mathrm{n,i}} < 1$ and $\sum_i \mathrm{w}^{\mathcal{G}}_{\mathrm{n,i}} = 1$) measures the similarity between the RGB intensities of patches $\boldsymbol{p_n}$ and $\boldsymbol{p_i}$ centered around the n-th and i-th pixels. Each patch is a set of pixels in a $R \times R$ window, so each $\mathbf{p}$ is a $3R^2$-dimensional ($R \times R \times RGB$) vector. It is a decreasing function of the Euclidean distance between the spatial RGB values as:

$$\mathrm{w}^{\mathcal{G}}_{\mathrm{n,i}} = \begin{cases} \dfrac{1}{\mathrm{z}^{\mathcal{G}}_{\mathrm{n}}} \exp^{-\frac{\|p_i - p_n\|^2}{\mathrm{h}^{\mathcal{G}}}}, & (\mathrm{n,i}) \in \mathrm{g_p}, \forall \mathrm{p} \\ 0, & \text{others} \end{cases} \tag{11}$$

where $\mathrm{z}^{\mathcal{G}}_{\mathrm{n}}$ is a normalization factor defined as $\mathrm{z}^{\mathcal{G}}_{\mathrm{n}} = \sum_{i \in g_p} \exp^{-\frac{\|p_i - p_n\|^2}{\mathrm{h}^{\mathcal{G}}}}$ to guarantee and ensure that $\sum_{i \in g_p} \mathrm{w}^{\mathcal{G}}_{\mathrm{n,i}} = 1$ and $\mathrm{h}^{\mathcal{G}}$ are a smoothing kernel for $3R^2$-dimensional vectors. The local-spectral weight $\mathrm{w}^{\mathcal{L}}_{\mathrm{n,j}}$ is defined in the exactly same format but with $\boldsymbol{p_n}$ and $\boldsymbol{p_i}$ being the RGB values of the n-th and i-th pixels (so each $\boldsymbol{p}$ is a three-dimensional vector here) and a smoothing kernel $h^{\mathcal{L}}$ for three-dimensional vectors.

We then build affinity matrices $\mathbf{W}^{\mathcal{G}} \in \mathbb{R}^{\mathrm{N \times N}}$ and $\mathbf{W}^{\mathcal{L}} \in \mathbb{R}^{N \times N}$, whose element encodes the pairwise similarity calculated using Eq. (11). Finally, the regularization term constrained by two types of self-similarities is represented as:

$$\boldsymbol{\Omega}(\mathbf{A}) = \left\|\mathbf{A} - \gamma \mathbf{W}^{\mathcal{G}} \mathbf{A} - (\mathbf{1} - \gamma) \mathbf{W}^{\mathcal{L}} \mathbf{A}\right\|_F^2 \tag{12}$$

With the self-similarity constraints of the global-structure and local-spectral, the sparse representation will be more robust and prospected to be similar for the locations in the same clustered global group and local superpixel. Given the HS dictionary $\mathbf{B}^*$ pre-learned using Eq. (9) and the regularization term with self-similarity in Eq. (12), Eq. (8), is convex and can be efficiently solved by optimization algorithm. We apply the ADMM technique to solve Eq. (8), and please refer to [45] for detail optimization procedure.

3.3 Experimental results

We evaluate the self-similarity constrained sparse representation method using two publicly released hyperspectral imaging databases: the CAVE and Harvard datasets. The CAVE dataset includes 32 indoor images consisting of paintings, toys, food, and so on, which are captured under controlled illumination. The Harvard dataset has 50 indoor and outdoor images captured under daylight illumination. The image size in the CAVE dataset is 512 × 512 pixels, and 31 spectral bands of 10 nm wide, which covers the visible spectrum from 400 to 700 nm. The image size in the Harvard dataset is 1392 × 1040 pixels, and 31 spectral bands of width 10 nm, basically covering the visible spectrum from 420 to 720 nm. In our experiments, we extract the top left 1024 × 1024 pixels as the understudying HR images. We take the original images in the datasets as ground-truth $\mathbf{Z}$ and resize them by a factor of 32 to create 16 × 16 images in the CAVE dataset and 32 × 32 images in the Harvard dataset, which is implemented by averaging over 32 × 32 pixel blocks as done in [10, 21]. The observed HR-RGB images Y are generated by multiplying the spectral channels of the ground-truth image with the spectral response R of a Nikon D700 camera. We evaluate the recovery performance of the estimated HS images using four quantitative metrics including root-mean-square error (RMSE), peak-signal-to-noise ratio (PSNR), spectral angle mapper (SAM) [9], and relative dimensionless global error in synthesis (ERGAS) [34]. The quantitative metric, SAM [9], gives the spectral distortion degree of the pixel spectrum in the estimated HR-HS image with the corresponding one in the ground-truth HR-HS image. We calculate the overall SAM metric of one understudying by averaging the SAMs computed from all pixels. The value of SAM is expressed in degrees and thus normalized into the range (−90, 90). The smaller the absolute value of SAM, the less the spectral distortion is. The ERGAS [34] calculates the average amount of the relative difference error, where the absolute difference error is normalized by intensity mean in each band. The smaller the ERGAS, the smaller the relative difference error is.

3.3.1 Compare results with the state-of-the-art methods

Firstly, we manifest the compared recovery performance of the HR-HS images with our proposed method (including the online dictionary learning procedure and self-similarity constraints) and the state-of-the-art HS image SR methods including matrix factorization (MF) method [18], coupled nonnegative matrix factorization method [19], sparse nonnegative matrix factorization (SNNMF) method [20], generalization of simultaneous orthogonal matching pursuit method [13], Bayesian sparse representation (BSR) method [9], couple spectral unmixing (CSU) method [10], and nonnegative structured sparse representation method [21]. **Table 1** manifests the average RMSE, PSNR, SAM, and ERGAS results of the 32 images in the CAVE dataset [32], while **Table 2** shows the average results of the 50 images from the Harvard dataset [33].

It can be seen from **Tables 1** and **2** that our approach obtains the best recovery performance for all quantitative metrics, and the performance improvement on the

	MF [18]	CNMF [19]	SNMF [20]	GSOMP [13]	BSR [9]	CSU [10]	NNSR [21]	Our
RMSE	3.03 ± 0.97	2.93 ± 1.30	3.26 ± 1.57	6.47 ± 2.53	3.13 ± 1.57	3.0 ± 1.40	2.21 ± 1.19	**2.17** ± **1.08**
PSNR	39.37 ± 3.76	39.53 ± 3.55	38.73 ± 3.79	32.48 ± 3.08	39.16 ± 3.91	39.50 ± 3.63	42.26 ± 4.11	**42.28** ± **3.86**
SAM	6.12 ± 2.17	5.48 ± 1.62	6.50 ± 2.32	14.19 ± 5.42	6.75 ± 2.37	5.8 ± 2.21	4.33 ± 1.37	**3.98** ± **1.27**
ERGAS	0.40 ± 0.22	0.39 ± 0.21	0.44 ± 0.23	0.77 ± 0.32	0.37 ± 0.22	0.41 ± 0.27	0.30 ± 0.18	**0.28** ± **0.18**

Table 1.
Quantitative comparison results of the self-similarity constrained sparse representation with the state-of-the-art methods on the CAVE dataset.

	MF [18]	CNMF [19]	SNMF [20]	GSOMP [13]	BSR [9]	CSU [10]	NNSR [21]	Our
RMSE	1.96 ± 0.97	2.08 ± 1.34	2.20 ± 0.94	4.08 ± 3.55	2.10 ± 1.60	1.7 ± 1.24	1.76 ± 0.79	**1.64** ± **1.20**
PSNR	43.19 ± 3.87	43.00 ± 4.44	42.03 ± 3.61	38.02 ± 5.71	43.11 ± 4.59	43.40 ± 4.10	44.00 ± 3.63	**45.20** ± **4.56**
SAM	2.93 ± 1.06	2.91 ± 1.18	3.17 ± 1.07	4.99 ± 2.99	2.93 ± 1.33	2.9 ± 1.05	2.64 ± 0.86	**2.63** ± **0.97**
ERGAS	0.23 ± 0.14	0.23 ± 0.11	0.26 ± 0.27	0.41 ± 0.24	0.24 ± 0.15	0.24 ± 0.20	0.21 ± 0.12	**0.16** ± **0.15**

Table 2.
Quantitative comparison results of the self-similarity constrained sparse representation with the state-of-the-art methods on the Harvard dataset.

CAVE dataset is more significant than on the Harvard dataset. The NNSR method [21] has the closest performance to ours, and both methods show relatively larger advantage over other methods. In addition, our method shows the best improvement on SAM values over NNSR [21]. This is because for SAM, a slight spectral distortion of the pixels with small magnitudes affects its value greatly. Thus, we can conclude that our proposed approach not only robustly recovers the HS image but also suppresses the noise and artifacts, especially for those pixels with small spectral magnitudes, due to the imposed constraints of the global-structure and local-spectral self-similarities.

3.3.2 Compared results without self-similarity constraints

One of the key differences of our method from existing ones (such as MF [18]) is the two types of imposed self-similarities formulated by the regularized term, $\mathbf{\Omega}(\mathbf{A})$ in Eq. (8). Without the $\mathbf{\Omega}(\mathbf{A})$ term, Eq. (8) can still be solved by an optimization method such as the ADMM. In addition, we can also adopt either global or local self-similarity separately, i.e., by taking only the $\mathbf{W}^{\mathcal{G}}$ or $\mathbf{W}^{\mathcal{L}}$ terms in Eq. (12). We conduct such experiments under the same experimental conditions, and the same quantitative metrics as in **Tables 1** and **2** for both datasets are shown in **Table 3**.

	CAVE dataset			Harvard dataset		
	Without both	**Local simil. only**	**Global simil. only**	**Without both**	**Local simil. only**	**Global simil. only**
RMSE	2.81 ± 1.42	2.25 ± 1.15	2.32 ± 1.20	1.83 ± 1.30	1.66 ± 1.20	1.88 ± 1.32
PSNR	40.05 ± 3.97	42.00 ± 3.91	41.78 ± 4.05	44.16 ± 4.39	45.01 ± 4.51	44.02 ± 4.56
SAM	5.46 ± 1.89	4.24 ± 1.36	4.59 ± 1.46	2.86 ± 1.06	2.69 ± 1.00	2.99 ± 1.09
ERGAS	0.37 ± 0.20	0.30 ± 0.18	0.31 ± 0.19	0.23 ± 0.16	0.19 ± 0.15	0.18 ± 0.16

Table 3.
Results without local, global, and both similarities on the CAVE and Harvard datasets.

Taking local self-similarity only into consideration significantly improves the results on both datasets for all quantitative metrics which shows relatively larger contribution than considering global self-similarity only, but integrating global self-similarity as our complete approach could further improve the results.

3.3.3 Evaluation results by changing parameter γ

In addition, we evaluate the HR-HS image recovery performance via changing the parameter γ for adjusting the contribution of global-structure and local-spectral self-similarity. For CAVE dataset, the parameter γ is changed from 0 (local-spectral self-similarity only) to 1 (global-structure self-similarity only) with interval 0.1, and apply the same measure metrics for manifesting the contribution of the global and local self-similarity. **Figure 3** (a)–(d) gives the curves of the quantitative measures, RMSE, PSNR, SAM, and ERGAS, respectively, which manifests that $\gamma = 0.3$ gives the best performances. For Harvard dataset, we also conducted experiments with the parameter γ, 0, 0.1, 0.2, ⋯, and the curves of the quantitative measures, RMSE, PSNR, SAM, and ERGAS, are given in **Figure 4**.

3.3.4 Visual quality comparison

Figures 5 and **6** manifest the recovered HS images and the difference images with respect to the ground-truth, which includes one example from the CAVE and Harvard dataset, respectively. Since including our method, the CSU [10] and NNSR [21] methods provide the impressive performance compared with all other evaluated methods as shown in **Tables 1** and **2**, we only give the compared results of our method, the CSU [10] and NNSR [21] methods for checking the differences in visual quality. It is obvious that the recovered HS images by our approach have smaller absolute difference magnitude for most pixels than the result by the CSU and NNSR method. It is also worth noting that when self-similarity is not applied,

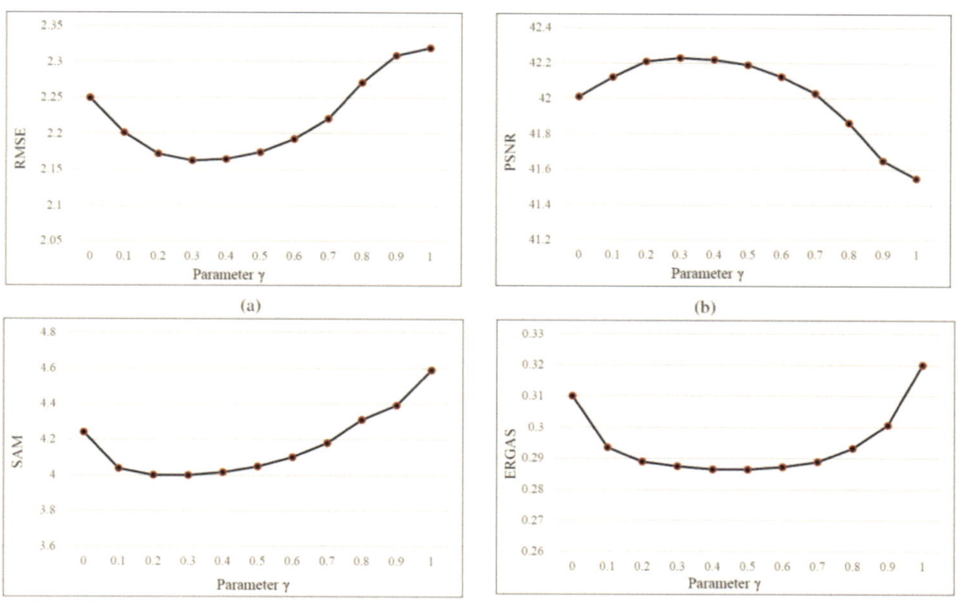

Figure 3.
The evaluated performances with different values of the parameter γ on CAVE dataset. (a) RMSE, (b) PNSR, (c) SAM, and (d) ERGAS.

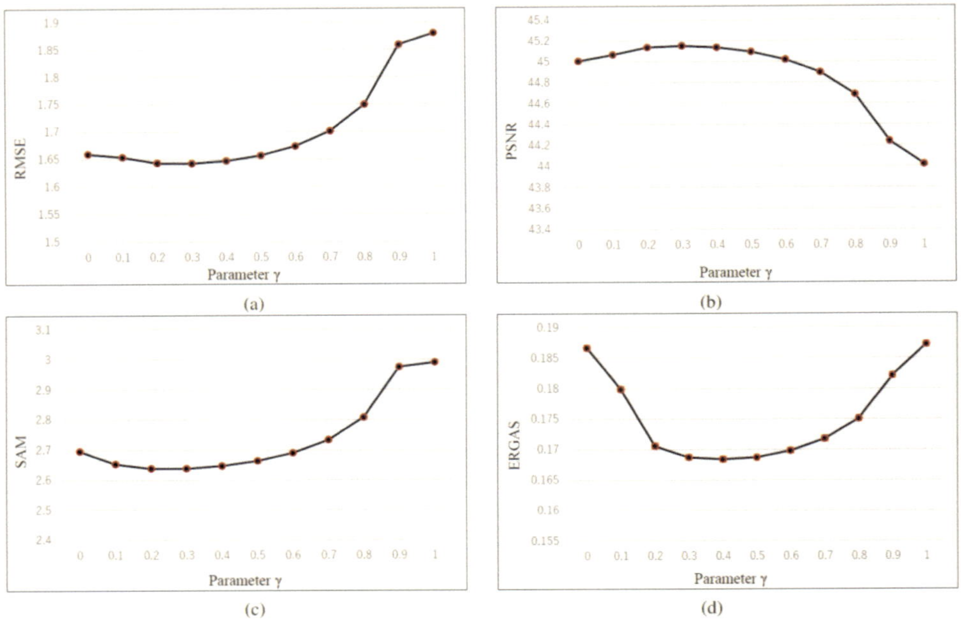

Figure 4.
The evaluated performances with different values of the parameter γ on Harvard dataset. (a) RMSE, (b) PNSR, (c) SAM, and (d) ERGAS.

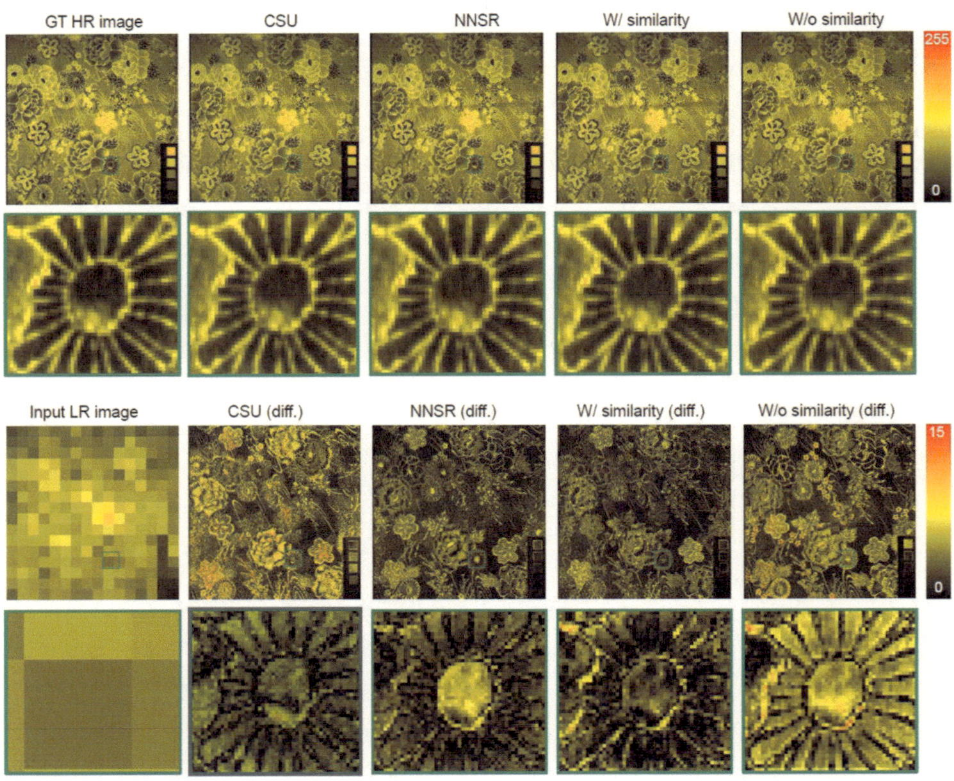

Figure 5.
The visualized results of the recovered HR images from the "cloth" image in the CAVE dataset. The first column shows the ground-truth HR image and the input LR image, respectively. The second to fifth columns show results from CSU [10], NNSR [21], and our method with and without self-similarity, where the upper part provides the recovered images and the lower part gives the absolute difference maps w.r.t. ground-truth. Close-up views are provided below each full resolution image.

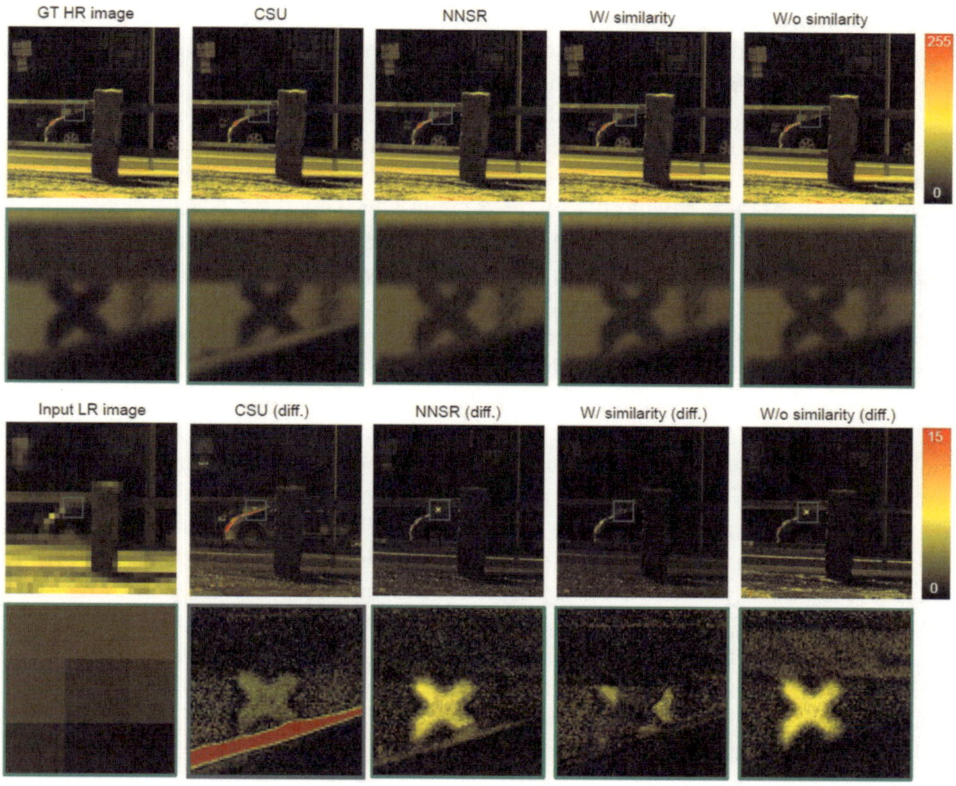

Figure 6.
The visualized results of the recovered HR images from the "imgf1" image in the Harvard dataset. The first column shows the ground-truth HR image and the input LR image, respectively. The second to fifth columns show results from CSU [10], NNSR [21], and our method with and without self-similarity, with the upper part showing the recovered images and the lower part showing the absolute difference maps w.r.t. ground-truth. Close-up views are provided below each full resolution image.

our results manifest quite similar appearance to those from the NNSR method [21], which also reflects the effectiveness of imposing the self-similarity constraint.

4. DCNN-based HS image super-resolution

Motivated by the success for image super-resolution and simply formulation, our previous work explored a simple DCNN-based HS image super-resolution method following the similar CNN structure as in [46], which mainly consists of three convolutional layers and was explained as three operations for the mapping process from LR images to HR images. This explanation follows the schematic concept in sparse coding-based SR: patch extraction, representation learning, nonlinear mapping, and reconstruction. Patch extraction obtains the overlapping patches from the input image and represents each patch as a high-dimensional vector. The convolution layers in CNN are used as feature learning and act as a nonlinear function, which maps a high-dimensional vector (conceptually the patch representation) to another high-dimensional vector (the feature map in the middle-layer of CNN). Reconstruction process combines the mapped CNN features into the final HR image. The above CNN architecture for Y-component recovery of natural image SR adopts the spatial filters in three convolutional layers with sizes $9 \times 9, 1 \times 1$, and 5×5. Since HSI SR attempts to recover high resolution in not only spatial but also spectral domain, which has been proven that the spectral response is more important in HIS SR, we set the spatial filter sizes as 3×3, 3×3, and 5×5 with full connection in spectral domain from either one of the available LR-HS and HR-RGB images or the concatenated LR-HS and HR-RGB cubic data.

The intuitive way to apply the above baseline architecture of CNN for HSI SR is to learn the HR-HS image, **Z** directly from the available LR-HS image **X**, called as spatial CNN. Another research line exploits CNN architecture for learning HSI SR **Z** from the available HR-MS (RGB) image **X**, named as spectral CNN. However, spatial CNN and spectral CNN take only one domain data of the available LR-HS and HR-MS images, **X** or **Y** as input, and completely exclude the other domain data. Therefore, this chapter introduces a spatial and spectral fusion architecture of CNN, named as SSF-CNN for recovering the HR-HS image. Recent CNN work incorporates shorter connections between layers for more accurate and efficient training of substantially deeper architectures such as ResNets and Highway Networks, or exploits concatenation between different layer for information and feature reuse such as Densenet, which manifest considerable improvements in different applications. In the scenario of our HSI SR application, since the available HR-RGB image has the same high spatial resolution and the expanding factor (about 10 from 3 to 31) in spectral domain is much smaller than those in spatial domain (32 times from 16/32 to 512/1024 in horizontal and vertical directions, respectively), we concatenate the available HR-RGB image (a part data of the input: Partial) to the outputs of the Conv and RELU blocks (Densely) in the CNN structure for transferring the available maximum spatial information, and name this new CNN architecture as PDCon-SSF. The schematic structures of the spatial CNN, spectral CNN, SSF-CNN and PDCon-SSF are shown in **Figure 7**.

Recently, we also investigated a residual network architecture for HS image super-resolution. The residual network takes the concatenated cubic data of both available HR-RGB and upsampled LR-HS images as input, and simultaneously maintains spectral attribute in LR-HS image and spatial context in HR-RGB image to estimate a more robust HS-HS image. Taking consideration of the characteristic in HS image super-resolution, we modified the ResNet architecture, which is originally proposed for solving higher-level computer vision problems such as image classification and detection, via removing unnecessary modules to simplify the network architecture for this low-level vision problem. Furthermore, as evidenced in pansharping research that the estimated HR-HS image should have similar spatial structure information with HR-RGB image, we utilize the input RGB image to guide the spatial structure of the learned feature maps in our proposed ResNet. We firstly upsample the LR-HS image to the same size with the HR-RGB image, and stack them together with a "Concat" layer in our method. Multiple residual layer modules with alternately conjuncted spectral and spatial reconstruction layers, which are implemented with convolutional kernel size 1 and n ($n > 1$), are used for effectively investigating the nonlinear spectral mapping and spatial structure. Our constructed ResNet architecture consists of 5 residual blocks and each block includes a set of the conjuncted spectral and spatial reconstruction layers as shown in **Figure 8**. In **Figure 8**, the first 3 residual blocks have 128 feature maps, and the last 2 residual blocks are with 256 feature maps. The output of the m-th residual block is expressed as:

$$\boldsymbol{F}_m = \mathbf{Spat}^3\left(\mathbf{Spec}^1(\boldsymbol{F}_{m-1})\right) + \boldsymbol{F}_{m-1} \tag{13}$$

where $\mathbf{Spec}^1(\cdot)$ denotes the spectral reconstruction layer with convolutional kernel size 1, and $\mathbf{Spat}^3(\cdot)$ denotes the spatial reconstruction layer with convolutional kernel size 3. $\boldsymbol{F}_{m-1}$ is the input of the residual block. Furthermore, considering the HR spatial structure in the observed HR-RGB image, we use the HR-RGB image to guide the spatial structure of the learned feature maps in the

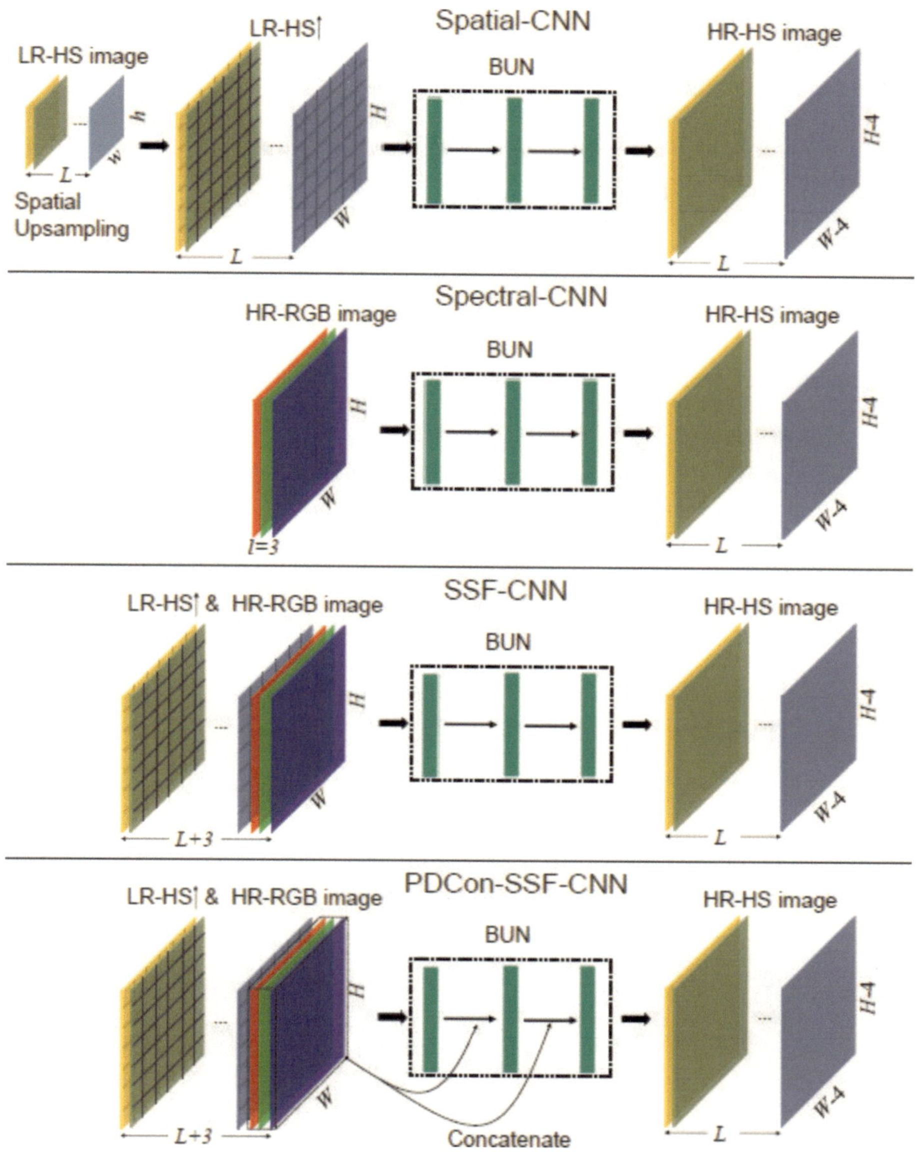

Figure 7.
The network architectures of four different types of CNNs. The top row denotes the baseline upsampling network, and the bottom rows are the architectures of spatial CNN, spectral CNN, and the SSF-CNN, respectively.

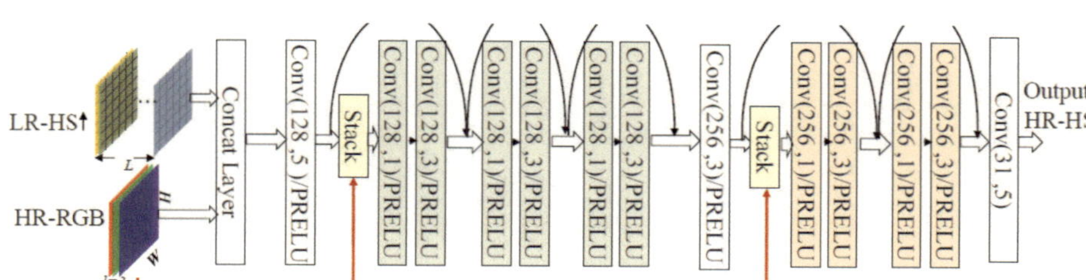

Figure 8.
The ResNet architecture for the residual component reconstruction.

residual blocks, which is modeled by stacking the input HR-RGB image and the input feature map F_{m-1}. Thus, with the added guidance connection, the output of a residual block is modified as:

$$F_m = \mathbf{Spat}^3(\mathbf{Spec}^1(\boldsymbol{stack}(F_{m-1}, Y))) + F_{m-1} \tag{14}$$

The guidance connections of the HR-RGB image are shown in dot lines in **Figure 7**. Our ResNet-based HR-HS image recovery model is trained by minimizing the Mean Square Error (MSE) between the estimated HR-HS image and the ground-truth **Z**.

4.1 Experimental results

We also validate the performance of the HR image reconstruction with the DCNN-based method using CAVE and Harvard datasets. We have randomly selected 20 HSIs from CAVE database to train CNN model, and the remainder is used for validation of the performance of the proposed CNN method. For Harvard database, 10 HSIs have been randomly selected for CNN model training, and the remainder 40 HSIs are as test for validation. **Figure 9** manifests the HR-RGB images of the test samples from CAVE database and several test samples from Harvard databases.

4.1.1 Compared results of different CNN models

As we introduced above, the CNN-based method can be used for recovering the HR-HS image from either of the available LR-HS, HR-RGB images or the concatenated cubic data of the LR-HS, HR-RGB images, which are named as spatial CNN, spectral CNN, Spatial and spectral Fusion CNN (SSF-CNN) and an extended

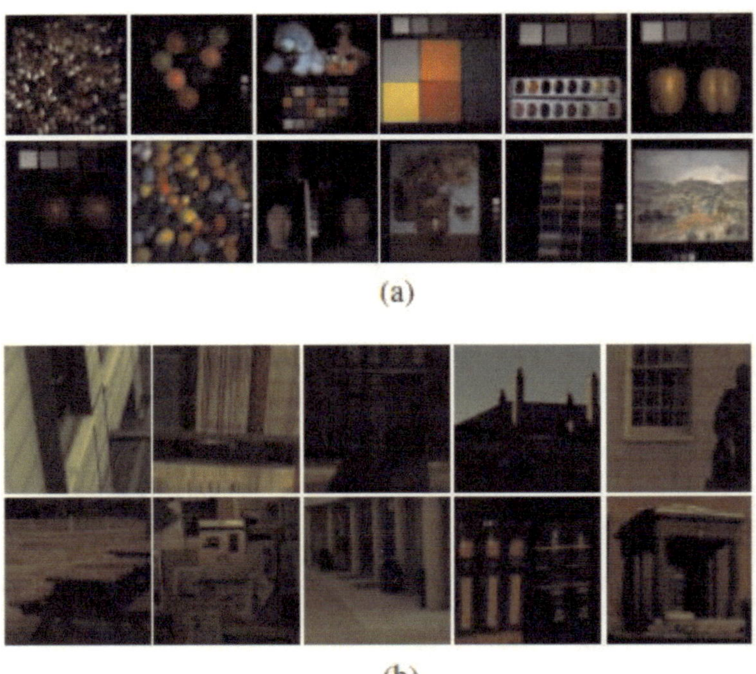

Figure 9.
The HR-RGB images of test samples from CAVE and Harvard databases.

version of SSF-CNN, PDCon-SSF. The baseline network is a three-layer convolution architecture. For CAVE database, we randomly select 20 images for learning the different types of CNN models, and save the CNN model parameters after 0.5 and 1 million iterations. The remainder 12 images in CAVE database are used for evaluating the recovering performance of different CNN models. The average and the standard deviation of RMSE, PSNR, SAM, and ERGAS of the 12 test images in CAVE database are shown in **Table 4**, which manifests much better results of the spectral CNN than spatial CNN due to the smaller expanding factor in spectral domain (about 10 from 3 to 32) than spatial domain (32 from 16 to 512 for horizontal and vertical directions, respectively) and significant performance improvement using SSF-CNN and PDCon-SSF-CNN models. One recovered HS image example and the corresponding residual images with the ground-truth HR images from CAVE database are visualized in **Figure 10** using different CNN models.

From **Table 4** and **Figure 10**, it can be seen that the SSF-based CNN models provide significant performance improvement compared with the spatial CNN and the spectral CNN, and thus for Harvard database, we only train the SSF-CNN and PDCon-SSF models with 1 million iterations using 10 randomly selected 10 images, and the remainder 40 images are used for evaluation. In addition, in order to validate the generation of the learned CNN model, we predict the HR-HS image of the Harvard test samples according to the parameters of the learned SSF-CNN and

CNN Models	Spatial-CNN	Spectral-CNN	SSF-CNN	PDCon-SSF
	0.5 million iterations			
RMSE	22.27±8.95	6.29±2.96	2.68±0.87	2.52±0.83
PSNR	21.66±3.58	32.89±3.60	40.09±3.25	40.59±3.24
SAM	19.53±11.01	9.60±3.56	5.04±1.47	4.95±1.54
ERGAS	2.37±0.83	0.60±0.21	0.28±0.10	0.27±0.11
	1 million iterations			
RMSE	24.49±9.35	5.73±2.20	2.20±0.73	**2.18±0.75**
PSNR	20.97±3.53	33.63±3.70	41.85±3.48	**41.93±3.57**
SAM	17.61±6.30	8.40±2.37	4.39±1.42	**4.38±1.39**
ERGAS	2.54±0.89	0.57±0.37	0.23±0.11	**0.22±0.10**

Table 4.
The average and standard deviation of RMSE, PSNR, SAM, and ERGAS using different CNN models of three-layer architecture under 0.5 and 1 million iteration training on CAVE database.

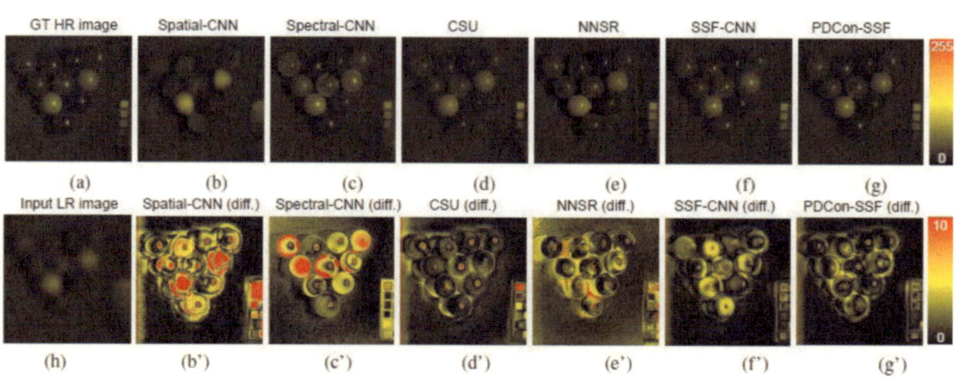

Figure 10.
The "superballs" image example from the CAVE database. The first row shows the ground-truth HR image and the recovered images by spatial CNN, spectral CNN, CSU [22], NNSR [12], and the proposed spatial and spectral CNN architectures, SSF-CNN and PDCon-SSF-CNN, respectively. The second row gives the input LR image, the absolute difference images between the ground-truth image, and the recovered HR-HS images in the first row.

PDCon-SSF-CNN with the CAVE training samples. The average and the standard deviation of RMSE, PSNR, SAM, and ERGAS of the 40 test images in Harvard database are shown in **Table 5**, which shows that the learned SSF-CNN and PDCon-SSF models even with the training samples from CAVE database can provide reasonable recovery performance and the quantitative measures can further be improved using the learned SSF-CNN and PDCon-SSF models even with 10 training images only. One recovered HS image example and its corresponding residual images with the ground-truth HR image from Harvard database are visualized in **Figure 11** using the learned SSF and PDCon-SSF-CNN models with the CAVE and Harvard training samples, respectively.

4.1.2 Compared results of different baseline CNN architectures

As mentioned above, we also investigated a residual network architecture for HS image super-resolution, which has different baseline CNN architecture with the SSF-CNN. Under the same experimental results, we implemented the DCNN-based HS image reconstruction using three-layer CNN and the ResNet architecture with five residual blocks. The compared quantitative results are shown in **Table 6** for both CAVE and Harvard datasets. One recovered HS image example and the corresponding residual images with the ground-truth HR image from CAVE database are visualized in **Figure 12** using the ResNet-RGB, SSF-Net, and the ResNet-based fusion models.

CNN Models	SSF-CNN-CAVE	PDCon-SSF-CAVE	SSF-CNN	PDCon-CNN
RMSE	4.23±2.36	3.97±2.08	1.76±1.00	**1.74±0.95**
PSNR	36.88±4.81	37.31±5.58	44.31±4.28	**44.33±4.18**
SAM	5.43±1.28	5.36±1.24	3.05±1.12	**3.03±1.15**
ERGAS	1.79±0.63	1.37±0.521	0.22±0.18	**0.20±0.17**

Table 5.
The average and standard deviation of RMSE, PSNR, SAM, and ERGAS of the test samples of Harvard database using different CNN models, where "SSF-CNN-CAVE" and "PDCon-SSF-CAVE" denote the learned CNN models using the training images from CAVE database.

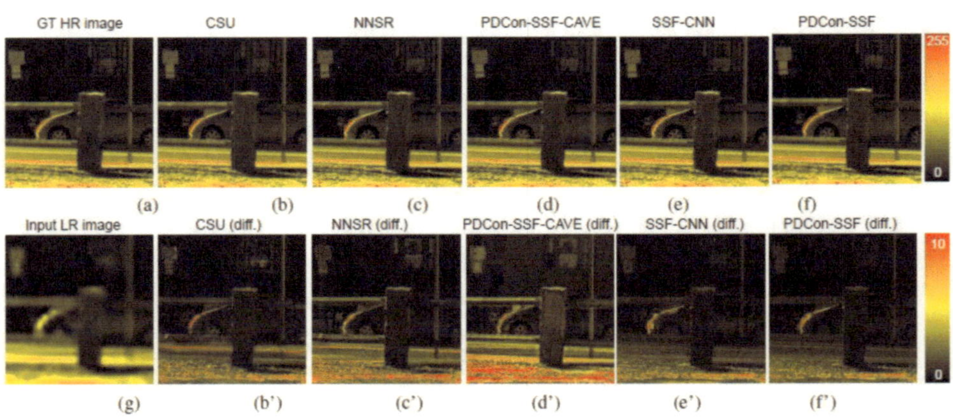

Figure 11.
An image example from the Harvard database. The first row shows the ground-truth HR image and the recovered images by CSU [22], NNSR [12], and the proposed PDCon-SSF-CNN using CAVE training images, SSF-CNN, and PDCon-SSF-CNN using Harvard training images, respectively. The second row gives the input LR image, the absolute difference images between the ground-truth image, and the recovered HR-HS images in the first row.

(a)

Inputs	ResNet-RGB	SSF-Net [20]	Ours
RMSE	4.77±2.44	2.20±0.73	**1.93±1.13**
PSNR	35.51±4.23	41.85±4.48	**43.57±4.64**
SAM	8.32±2.67	4.39±1.42	**3.58±1.40**
SSIM	0.977±0.02	0.986±0.01	**0.991±0.01**

(b)

Inputs	ResNet-RGB	SSF-Net [20]	Ours
RMSE	3.00±2.26	1.94±1.23	**1.83±1.01**
PSNR	40.40±5.77	43.56±4.88	**44.05±4.85**
SAM	3.83±1.54	3.14±0.97	**2.37±0.65**
SSIM	0.980±0.01	0.988±0.01	**0.984±0.01**

Table 6.
The compared average and standard deviation of RMSE, PSNR, SAM, and SSIM using the ResNet-RGB, SSF-Net [51], and the ResNet-based fusion methods on both CAVE and Harvard databases.

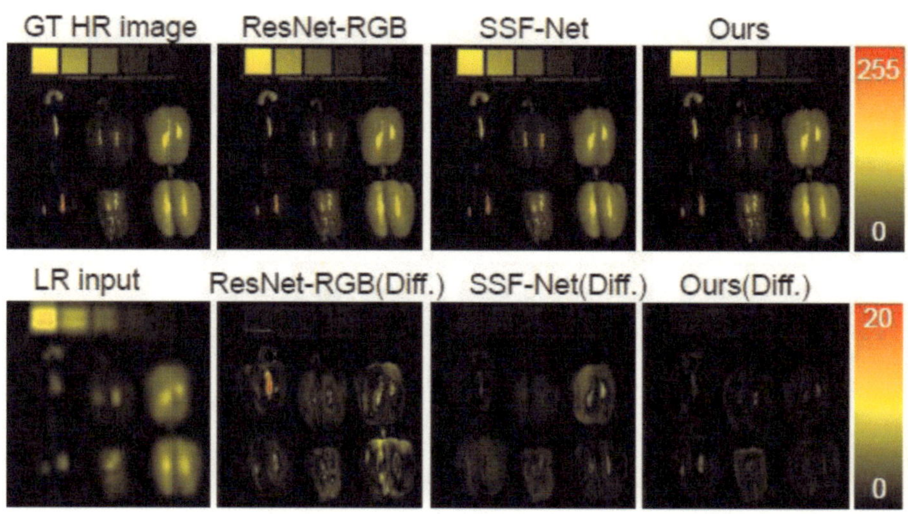

Figure 12.
The visualized results of the recovered HR images from an example image in CAVE dataset.

5. Conclusions

This chapter introduced recently research on HS image super-resolution. We firstly described the problem formulation for HS image super-resolution and provided the mathematical model between the observed HR-RGB, LR-HS images, and the required HR-HS image. Then we gave the detail description for an optimization-based method: self-similarity constrained sparse representation and the recently proposed DCNN-based method. Experimental results validated that the recently proposed HR image super-resolution methods manifest promising performance on benchmark datasets.

Author details

Xian-Hua Han
Graduate School of Science and Technology for Innovation, Yamaguchi University, Yamaguchi, Japan

*Address all correspondence to: hanxhua@yamaguchi-u.ac.jp

References

[1] Fauvel M, Tarabalka Y, Benediktsson J, Chanusssot J, Tilton J. Advances in spectral-spatial classification of hyperspectral images. Proceedings of the IEEE. 2013;**101**(3): 652-675

[2] Uzair M, Mahmood A, Mian A. Hyperspectral face recognition using 3d-dct and partial least squares. BMVC. 2013:57.1-57.10

[3] Zhang D, Zuo W, Yue F. A comparative study of palmprint recognition algorithm. ACM Computing Surveys. 2012;**44**(1):2:1-2:37

[4] Nguyen H, Benerjee A, Chellappa R. Tracking via object reflectance using a hyperspectral video camera. CVPRW. 2010:44-51

[5] Tarabalka Y, Chanusssot J, Benediktsson J. Segmentation and classification of hyperspectral images using minimum spanning forest grown from automatically selected markers. IEEE Transactions on Systems, Man, and Cybernetics: Systems. 2010;**40**(5): 1267-1279

[6] Zhou Y, Chang H, Barner K, Spellman P, Parvin B. Classification of histology sections via multispectral convolutional sparse coding. CVPR. 2014:3081-3088

[7] Bioucas-Dias J, Plaza A, Camps-Valls G, Scheunders P, Nasrabadi NM, Chanussot J. Hyperspectral remote sensing data analysis and future challenges. IEEE Geoscience and Remote Sensing Magazine. 2013; **1**(2):6-36

[8] Akhtar N, Shafait F, A M. Sungp: A greedy sparse approximation algorithm for hyperspectral unmixing. ICPR. 2014: 3726-3731

[9] Wei Q, Bioucas-Dias J, Dobigeon N, Toureret J. Hyperspectral and multispectral image fusion based on a sparse representation. IEEE Transactions on Geoscience and Remote Sensing. 2015;**53**(7):3658-3668

[10] Lanaras C, Baltsavias E, Schindler K. Hyperspectral superresolution by coupled spectral unmixing. ICCV. 2015: 3586-3595

[11] Grohnfeldt C, Zhu XX, Bamler R. Jointly sparse fusion of hyperspectral and multispectral imagery. IGARSS. 2013:4090-4093

[12] Akhtar N, Shafait F, Mian A. Bayesian sparse representation for hyperspectral image super resolution. CVPR. 2015:3631-3640

[13] Akhtar N, Shafait F, Mian A. Sparse spatio-spectral representation for hyperspectral image super-resolution. ECCV. 2014:63-78

[14] Elad M, Aharon M. Image denoising via sparse and redundant representations over learned dictionaries. IEEE Transactions on Image Processing. 2006;**15**(12): 3736-3745

[15] Tropp JA, Gilbert AC. Signal recovery from random measurements via orthogonal matching pursuit. IEEE Transactions on Information Theory. 2007;**53**(12):4655-4666

[16] Donoho DL, Tsaig Y, Drori I, Starck J-L. Sparse solution of underdetermined linear equations by stagewise orthogonal matching pursuit. IEEE Transactions on Information Theory. 2012;**58**(2):1094-1121

[17] Wright JA, Yang AY, Ganesh A, Sastry SS, Ma Y. Robust face recognition via sparse representation. IEEE Transactions on Pattern Analysis and Machine Intelligence. 2009;**31**(2): 210-227

[18] Kawakami R, Wright J, Tai Y-W, Matsushita Y, Ben-Ezra M, Ikeuchi K. High-resolution hyperspectral imaging via matrix factorization. CVPR. 2011: 2329-2336

[19] Yokoya N, Yairi T, Iwasaki A. Coupled nonnegative matrix factorization for hyperspectral and multispectral data fusion. IEEE Transactions on Geoscience and Remote Sensing. 2012;**50**(2):528-537

[20] Wcoff E, Chan T, Jia K, Ma W, Ma Y. A non-negative sparse promoting algorithm for high resolution hyperspectral imaging. ICASSP. 2013: 1409-1413

[21] Dong W, Fu F, Shi G, Cao X, Wu J, Li G, et al. Hyperspectral image super-resolution via non-negative structured sparse representation. IEEE Transactions on Image Processing. 2016; **25**(3):2337-2352

[22] Huang B, song H, Cui H, Peng J, Xu Z. Spatial and spectral image fusion using sparse matrix factorization. IEEE Transactions on Geoscience and Remote Sensing. 2014;**52**(3):1693-1704

[23] Chavez P, Sides S, Anderson J. Comparison of three different methods to merge multiresolution and multispectral data: Landsat tm and spot panchromatic. Photogrammetric Engineering and Remote Sensing. 1991; **30**(7):1779-1804

[24] Haydn R, Dalke G, Henkel J, Bare J. Application of the ihs color transform to the processing of multisensor data and image enhancement. International Symposium on Remote Sensing of Environment. 1982:559-616

[25] Aiazzi B, Baronti S, Lotti F, Selva M. A comparison between global and context-adaptive pansharpening of multispectral images. IEEE Geoscience and Remote Sensing Letters. 2009;**6**(2): 302-306

[26] Minghelli-Roman A, Polidori L, Mathieu-Blanc S, Loubersac L, Cauneau F. Spatial resolution improvement by merging meris-etm images for coastal water monitoring. IEEE Geoscience and Remote Sensing Letters. 2006;**3**(2):227-231

[27] Zurita-Milla R, Clevers J, Schaepman ME. Unmixing-based landsat tm ane meris fr data fusion. IEEE Geoscience and Remote Sensing Letters. 2008;**5**(3):453-457

[28] Cetin M, Musaoglu N. Merfing hyperspectral and panchromatic image data: Qualitative and quantitative analysis. International Journal of Remote Sensing. 2009;**30**(7):1779-1804

[29] Lee DD, Seung SH. Algorithms for non-negative matrix factorization. NIPS. 2001:556-562

[30] Bioucas-Dias JM, Plaza A, Dobigeon N, Parente M, Du Q, Gader P, et al. Hyperspectral unmixing overview: Geometrical, statistical and sparse regression-based approaches. IEEE Journal of Selected Topics in Applied Earth Observation and Remote Sensing. 2012;**5**(2):354-379

[31] Achanta R, Shaji A, Smith K, Lucchi A, Fua P, Susstrunk S. Slic superpixels compared to state-of-the-art superpixel methods. IEEE Transactions on Pattern Analysis and Machine Intelligence. 2012;**34**(11):2274-2282

[32] Yasuma F, Mitsunaga T, Iso D, Nayar S. Generalized assorted pixel camera: Post-capture control of resolution, dynamic range and spectrum. IEEE Transactions on Image Processing. 2010;**19**(9):2241-2253

[33] Chakrabarti A, Zickler T. Statistics of real-world hyperspectral images. CVPR. 2011:193-200

[34] Wald L. Quality of high resolution synthesised images: Is there a simple

criterion? Proceedings of Fusion Earth Data. 2000:99-103

[35] Minghelli-Roman A, Polidori L, Mathieu-Blanc S, Loubersac L, Cauneau F. Spatial resolution improvement by merging meris-etm images for coastal water monitoring. IEEE Geoscience and Remote Sensing Letters. 2006;**3**(2):227-231

[36] Zurita-Milla R, Clevers J, Schaepman ME. Unmixing-based landsat tm and meris fr data fusion. IEEE Geoscience and Remote Sensing Letters. 2008;**5**(3):453-457

[37] Duran J, Buades A, Sbert C, Blanchet G. A Survey of Pansharpening Methods with A New Band-Decoupled Variational Model. CoRR, vol. abs/1606.05703; 2016

[38] Kidiyo K, Miloud CE-M, Nasreddine T. Recent trends in satellite image pan-sharpening techniques. In: 1st International Conference on Electrical, Electronic and Computing Engineering. 2014

[39] Cetin M, Musaoglu N. Merging hyperspectral and panchromatic image data: Qualitative and quantitative analysis. International Journal of Remote Sensing. 2009;**30**(7):1779-1804

[40] Halimi A, Bioucas-Dias J, Dobigeon N, Buller G, McLaughlin S. Fast hyperspectral unmixing in presence of nonlinearity or mismodelling effects. Transactions on Computational Imaging. 2017;**3**(2):146-159

[41] Sigurdsson J, Ulfarsson M, Sveinsson J, Bioucas-Dias J. Sparse distributed multitemporal hyperspectral unmixing. IEEE Transactions on Geoscience and Remote Sensing. 2017; **55**(11):6069-6084

[42] Wei Q, Bioucas-Dias J, Dobigeon N, Tourneret J-Y, Chen M, Godsill SS. Multi-band image fusion based on spectral unmixing. IEEE Transactions on Image Processing. 2016;**54**(12):7236-7249

[43] Fu X, Ma W-K, Bioucas-Dias J, Chan T-H. Semiblind hyperspectral unmixing in the presence of spectral library mismatches. IEEE Transactions on Geoscience and Remote Sensing. 2016;**54**(9):5171-5184

[44] Lee DD, Seung SH. Algorithms for non-negative matrix factorization. NIPS. 2001:556-562

[45] Han X-H, Shi B, Zheng Y. Self-similarity constrained sparse representation for hyperspectral image superresolution. IEEE Transactions on Image Processing. 2018;**27**(11):5625-5637

[46] Dong C, Loy CC, He KM, Tang XO. Image super-resolution using deep convolutional networks. IEEE Transactions on Pattern Analysis and Machine Intelligence (TPAMI). 2015; **38**(2):295-307

[47] Kim J, Lee JK, Lee KM. Accurate image super-resolution using very deep convolutional networks. IEEE Conference on Computer Vision and Pattern Recognition (CVPR). 2016: 1646-1654

[48] Li YS, Hua J, Zhao X, Xie WY, Li JJ. Hyperspectral image super-resolution using deep convolutional neural network. Neurocomputing. 2017;**266**: 29-41

[49] Alvarez-Gila A, van de Weijer J, Garrote E. Adversarial networks for spatial contextaware spectral image reconstruction from rgb. In: IEEE International Conference on Computer VisionWorkshop (ICCVW 2017). 2017

[50] Galliani S, Lanaras C, Marmanis D, Baltsavias E, and Schindler K. Learned Spectral Super-Resolution," arXiv preprint arXiv:1703.09470; 2017

[51] Han X-H, Shi B, Zheng Y. SSF-CNN: Spatial and spectral fusion with cnn for hyperspectral image super-resolution. ICIP. 2018:2506-2510

Chapter 4

Fast Chaotic Encryption for Hyperspectral Images

Carlos Villaseñor, Javier Gomez-Avila, Nancy Arana-Daniel, Alma Y. Alanis and Carlos Lopez-Franco

Abstract

The information collected by hyperspectral images (HI) is essential in applications of remote sensing like object detection, geological process recognition, and identifying materials. However, HI information could be sensitive, and therefore, it should be protected. In this chapter, we show a parallel encryption algorithm specifically designed for HI. The algorithm uses multiple chaotic systems to produce a crossed multidimensional chaotic map for encrypting the image; the scheme takes advantage of the multidimensional nature of HI and is highly parallelizable, which leads to a time-efficient algorithm. We also show that the algorithm gets high-entropy ciphertext and is robust to ciphertext-only attacks.

Keywords: chaotic encryption, hyperspectral images, parallel computing

1. Introduction

Hyperspectral images (HI) or image spectrometry is a spectral sensing technique in which an object is photographed using several well-defined optical bands in the broad spectral range. This technique integrates imaging and spectroscopy to attain both spatial and spectral information from an object. It was originally developed on satellite and airborne platforms for remote sensing applications utilizing satellite imaging data of the earth and planets mostly; however, it has found application in diverse fields such as military defense, medical diagnosis, and agriculture [1].

HI are characterized by their spatial and spectral resolution. The spatial resolution measures the geometric relationship of the image pixels, and the spectral resolution determines the variations within image pixels as a function of wavelength. The HI has two spatial dimensions (m and n) and one spectral dimension (l). The hyperspectral data are represented in the form of a 3D hyperspectral data cube as it is shown in **Figure 1**. Each slice of the cube along the spectral dimension is called band or channel.

HI could be made up of hundreds of contiguous bands for each spatial position. Consequently, each pixel in a hyperspectral image contains a spectrum representing the light-absorbing and/or scattering properties of the spatial region represented by that pixel. The resulting spectrum acts like a signature, which can be used to classify or estimate composition of the material it represents.

However, the information in the HI could be sensitive, and thus, it should be protected. There already exist many algorithms for secure encryption [2], like advanced encryption standard (AES) [3–5] based on irreducible polynomials in

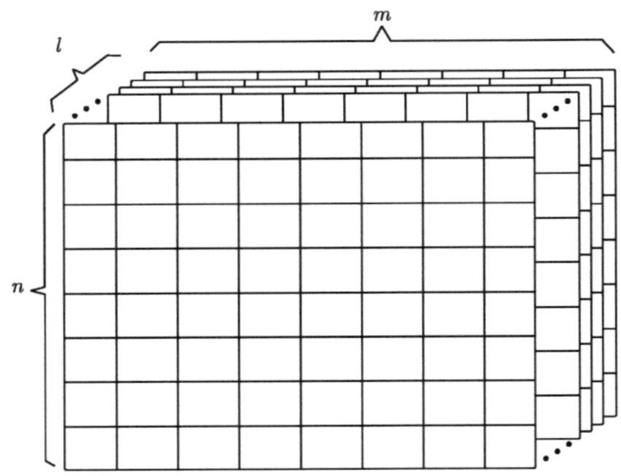

Figure 1.
Hyperspectral image dimensions.

Galois fields or the Rivest-Shamir-Adleman (RSA) algorithm [6] based on large prime number factorization. These algorithms work with raw data, and they offer a secure way to protect it. They work at the binary level of the data, and then they can deal with all kinds of data structures like images, videos, and documents.

The computational run-time of AES and RSA could be very high for a high volume of data. For this reason, in the last decades, the uses of chaos-based cryptography have become very popular [7]. The chaotic systems are deterministic dynamical systems which are sensitive to their initial conditions and parameters.

In this chapter, we present an extended revision of the parallel encryption algorithm based on chaotic systems for HI previously presented in [8]. This algorithm was specifically designed for HI, and it takes advantage of the spatial and spectral distribution to get fast encryption. The algorithm gets high-entropy ciphertext, and it is robust to ciphertext-only attacks.

The chapter is organized as follow: in Section 2, we briefly review the basics of HI and their applications. In Section 3, we explain the characteristics of the chaotic systems and why they are a suitable mathematical tool for encryption. In Section 4, we present our encryption scheme with full details of implementation (we include source code in Matlab and CUDA languages). In Section 5, we present new experimentation and results, and finally, in Section 6, we present our conclusions and future work.

2. Hyperspectral images

A digital image is a rectangular array of n rows and m columns where every element of the array, also called pixel, has an intensity value. New cameras allow us to have multiple measures in the same pixel of different wavelengths, where we represent this information by adding a new dimension l called channels.

The pixels are a discrete quantified measure of light; the intensity represented is an integer proportional to the number of photons detected. In a simple image, the intensity is represented with 2^8 precision, where 0 is no photons detected and 255 is the pixel saturated. Other systems use 2^{12}, 2^{14}, or 2^{16} levels of intensity [9].

Color images have three channels to represent visible light (400–800 nm of wavelength). This kind of camera uses filter for red, green, and blue light. When we have more of three channels, we called it multivariate images or multispectral images, for example, an image of 21 channels.

A common convention is to call hyperspectral images to images with more than 100 channels [9]. Every pixel of a HI has its spectral representation of that position;

classical spectrophotometers use a light source, a filter system to disperse the light into their respective wavelengths, and a detection system. The HI has commonly a spectral resolution of 10–20 nm.

HI have many applications such as oil detection [10], earth remote sensing [11], vegetation and water studies [12, 13], archaeology and art conservation [14–16], finding minerals and other materials [17], medical diagnosis [18–20], food quality and safety control [18, 21, 22], and crime scene analysis and forensic traces [23, 24].

3. Chaotic systems and chaotic encryption

Chaotic systems are dynamical systems that for a certain range of initial conditions show chaotic behavior. This chaotic behavior has the following features:

- Sensitive to initial conditions: the chaotic system is exponentially sensitive to small changes in the initial conditions. These small changes could produce a big difference in the system output.

- Bounded: the states with chaotic behavior have bounded limits.

- Deterministic: a chaotic system does not have randomness, so if we have the initial conditions and parameters, we can simulate the system.

- Aperiodic: there is no periodicity in the chaotic behavior.

The chaotic behavior is a deterministic phenomenon that seems random for a viewer that does not know the initial conditions and parameters of the system; but like any deterministic system, we can obtain the same time-series output by simulating the system with the same initial condition and parameters.

These features make the chaotic system a suitable platform for encryption. We have two principal ways of implementing chaotic encryption:

- Chaotic encryption by synchronization phenomenon: in the case of continuous signals, they could be encrypted by adding a chaotic continuous signal. For the decrypt process, we synchronize another chaotic system throughout the synchronization phenomenon and control theory to subtract the chaotic signal [25, 26].

- Simulating discrete chaotic system: it consists in generating a chaotic mask and mixing with a discrete plaintext with an involution operation. For decrypt, we generate the same mask, and by properties of the involutions, we recover the original signal. In this chapter, we concentrate in the second alternative.

Many chaotic systems have been used for different encryption problems like:

- Stream ciphers: Lorenz system has been used as a stream cipher in [27].

- Image encryption: for image encryption is common in using chaotic maps like Ikeda chaotic map [28], logistic map, tent map, quadratic map, and Bernoulli map [29], a survey is presented in [30].

- Video encryption: the chaotic maps also are used to encrypt video in [31–33].

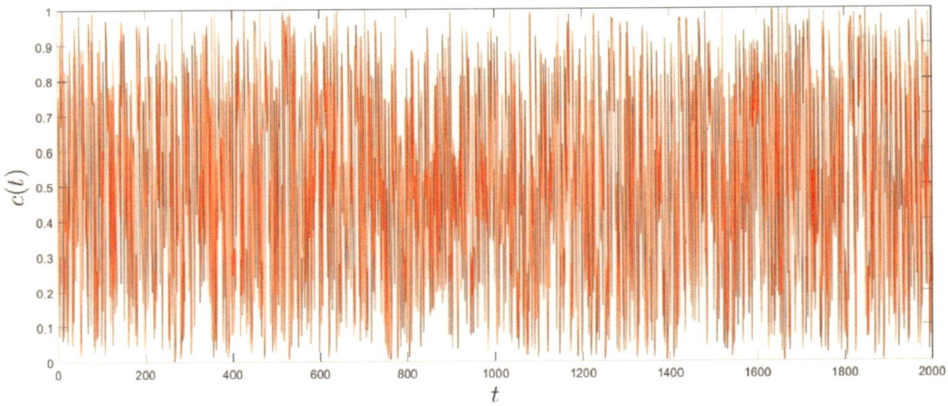

Figure 2.
Simulation of the PLM chaotic system.

For the proposed encryption scheme, we use the piecewise linear chaotic maps (PLM) described by Eq. (1), where $x_0 \in (0, 1)$ is the initial condition and $\mu \in (0, 0.5)$ is the parameter of the system. In **Figure 2**, we present a simulation of the PLM system.

$$x_{n+1} = \frac{x_n - \lfloor \frac{x_n}{\mu} \rfloor \mu}{\mu} \tag{1}$$

4. Chaotic systems for parallel encryption of hyperspectral images

In this section we will review the proposed chaotic encryption scheme for HI presented in [8]. The proposed scheme is shown in **Figure 3**. We propose to use four chaotic systems for encrypting HI. The first three chaotic systems are used to generate a chaotic mask of the same size of the HI, and the last one is used to generate a chaotic substitution box (S-box) according to the algorithm in [34].

The key is formed with initial conditions $\boldsymbol{x_i}$ and parameters $\boldsymbol{\mu_i}$ of the four PLM chaotic systems. It is important to notice that this scheme can be implemented with other discrete chaotic systems. These systems are simulated in parallel to create four different chaotic signals denoted by $\boldsymbol{c_1(t)}$, $c_2(t)$, $c_3(t)$, and $\boldsymbol{c_s(t)}$ that are saved as the key expansion.

In **Figure 4**, we show the source code of the parallel implementation of the chaotic system in the CUDA programming language, where the variable ite represents the maximum number of iterations of the chaotic system, x_vec is a vector with the initial condition parameters, mu_vec is a vector with the parameters of the PLM systems, and C is an array where the chaotic signals are stored.

The chaotic signal $c_s(t)$ is used to generate a chaotic S-box [34]. An S-box is a basic component for symmetric key ciphers because it is a reversible operation for confusion introduction. Confusion is a property of a secure cipher denied by Claude Shannon, and it means that each basic unit of the ciphertext should depend on several parts of the key hiding the relationship between them. The S-box is a permutation which is commonly implemented with a look-up table that maps from a byte to another byte.

In **Figure 5**, we present the Matlab code for the generation of the chaotic S-box and its inverse.

Using the previous codes, we can encrypt the HI by using Eq. (2), where every pixel in the direction (i, j, k) is encrypted in parallel by composing a chaotic signal with the $c_1(t), c_2(t), c_3(t)$ chaotic signals. We combine them by using the XOR

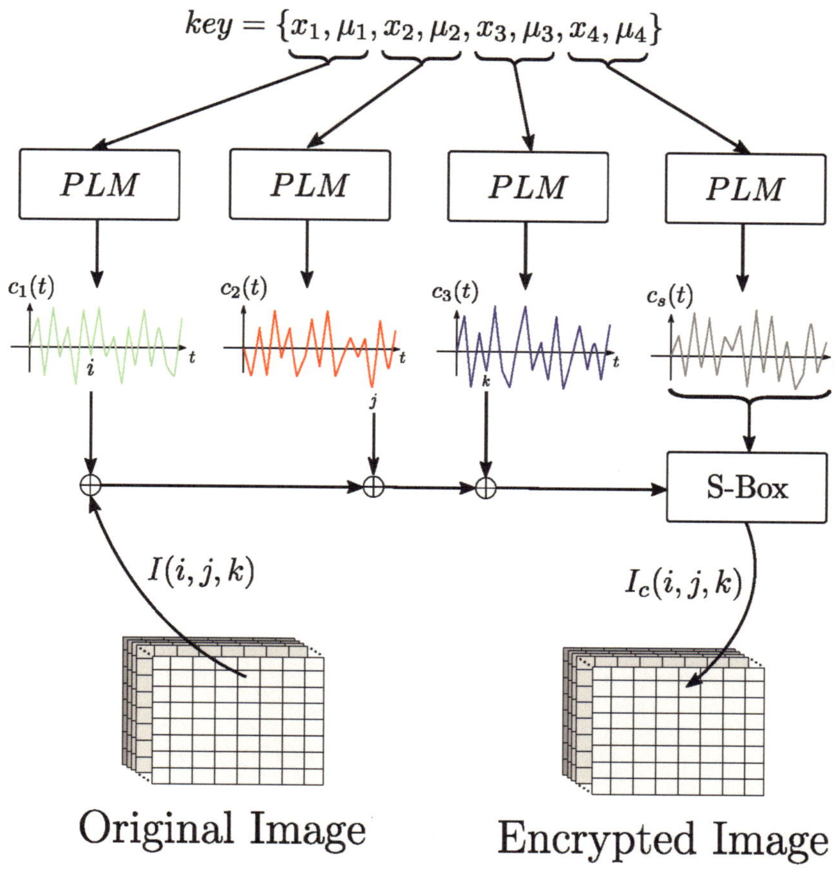

Figure 3.
Encryption scheme for HI.

```
#include "tmwtypes.h"

__device__ double PLM(double x, double mu)
{
        x = (x -floor)x / mu) * mu) / mu;
        return x;
}

__global__ void Chaos(uint8_T * C, const double* x_vec, const double* mu_vec, const
int ite)
{
        double temp_x = x_vec[threadIdx.x];
        for (int i = 0; i < ite; i++)
        {
                temp_x = PLM(temp_x, mu_vec[threadIdx.x]);
                C[ite*threadIdx.x + i] = (uint8_T)(255 * temp);
        }
}
```

Figure 4.
Parallel chaotic system implementation (Chaos.cu file).

operation which is an involution. Afterwards we use the chaotic S-box to include confusion in the signal:

$$I_c(i,j,k) = S_{box}[c_3(k) \bigoplus c_2(j) \bigoplus c_1(i) \bigoplus I(i,j,k)] \quad (2)$$

Eq. (2) is implemented in the CUDA language as it is shown in **Figure 6**.
To use the chaotic and encrypted function from Matlab, we compile from the terminal, in this case, using nvcc compiler with the lines in **Figure 7**:

```
function [S, invS] = Chaotic_SBox(x, mu)
    S = zeros(16*16,1);
    invS = zeros(16*16,1);
    Q = 0:255;
    max = 2^32;
    for i = 0:255
        x = PLM(x, mu);
        d = mod(floor(max*x), 256);
        j = mod(d,256 - i);
        S(i+1) = Q(j + 1);
        invS(Q(j+1) + 1) = i;
        Q(j+1) = Q(256-i);
    end
end
```

Figure 5.
Chaotic substitution box.

```
#include "tmwtypes.h"

__global__ void Encrypt(uint8_T * I, uint8_T * C, const int ite, uint8_T *S)
{
        // Alias
        int x = blockIdx.x;
        int y = threadIdx.y;
        int z = blockIdx.z;

        // Calculate image index
        int index = z * blockDim.y * gridDim.x + y *gridDim.x + x;

        // Encrypt
        I[index] = C[x] ^ C[ite + y] ^ C[2*ite + z] ^ I[index];
        I[index] = S[(int)I[index]];
}
```

Figure 6.
Parallel chaotic encryption (Encrypt.cu file).

```
nvcc -ptx Encrypt.cu -ccbin "C:\Program Files\Microsoft Visual Studio 12.0\VC\bin"
nvcc -ptx Chaos.cu -ccbin "C:\Program Files\Microsoft Visual Studio 12.0\VC\bin"
```

Figure 7.
Compilation of the CUDA files.

In **Figure 8**, we show the code in Matlab for encrypting HI in parallel. Notice that the first step in **Figure 8** is to load the HI in the variable I in the GPU memory (Igpu), and then we calculate the chaotic S-box in the variable S with the code in **Figure 5**; after that, we simulate the chaotic systems in the variable Cgpu with the code in **Figure 4**. Finally, we encrypt the HI with **Figure 6** gathering the result in the Ic variable.

The decryption scheme of the HI is almost symmetrical. This could be achieved with Eq. (3) which is implemented in **Figure 9**. The code in **Figure 9** is very close to the one in **Figure 6**, but notice that the inverse of the S-box is used and the XOR of the chaotic signal are in inverse order:

$$I(i,j,k) = c_1(i) \bigoplus c_2(j) \bigoplus c_3(k) \bigoplus S_{box}^{-1}[I_c(i,j,k)] \tag{3}$$

Using Eq. (2) in Eq. (3) we obtain the development in Eqs. (4)–(6) that shows how the S-box is canceled with the inverse S-box and how the chaotic signals are canceled because the XOR operation is an involution:

```
function Ic = Parallel_encryption(I, key)

    %% Size of hyperspectral image
    [i, j, k] = size(I);

    %% Load to GPU
    Igpu = gpuArray(I);

    %% Calculate Sustitution-Box
    [S, ~] = Chaotic_SBox(key(1), key(2));

    %% Parallel Simulation chaotic systems
    ite = max([i, j, k]);
    C = uint8(zeros(ite, 3));
    Cgpu = gpuArray(C);
    kernel = parallel.gpu.CUDAKernel('Chaos.ptx', 'Chaos.cu', 'Chaos');
    kernel.ThreadBlockSize = [3,1,1];
    Cgpu = feval(kernel, Cgpu, gpuArray([key(3); key(5); key(7)]),
                 gpuArray([key(4); key(6); key(8)]), ite);


    %% Parallel encryption
    kernel = parallel.gpu.CUDAKernel('Encrypt.ptx', 'Encrypt.cu', 'Encrypt');
    kernel.ThreadBlockSize = [1, j, 1];
    kernel.GridSize = [i, 1, k];
    result = feval(kernel, Igpu, Cgpu, ite, gpuArray(uint8(S)));

    %% Download to CPU
    Ic = gather(result);
end
```

Figure 8.
Encryption of the HI.

```
#include "tmwtypes.h"

__global__ void Decrypt(uint8_T * I, uint8_T * C, const int ite, uint8_T *invS)
{
        // Alias
        int x = blockIdx.x;
        int y = threadIdx.y;
        int z = blockIdx.z;

        // Calculate image index
        int index = z * blockDim.y * gridDim.x + y *gridDim.x + x;

        // Decrypt
        I[index] = invS[(int)I[index]];
        I[index] = C[2*ite + z] ^ C[ite + y] ^ C[x] ^ I[index];
}
```

Figure 9.
Parallel decryption function (Decrypt.cu).

$$I(i,j,k) = c_1(i) \bigoplus c_2(j) \bigoplus c_3(k) \bigoplus S_{box}^{-1}[S_{box}[c_3(k) \bigoplus c_2(j) \bigoplus c_1(i) \bigoplus I(i,j,k)]] \quad (4)$$

$$I(i,j,k) = c_1(i) \bigoplus c_2(j) \bigoplus c_3(k) \bigoplus c_3(k) \bigoplus c_2(j) \bigoplus c_1(i) \bigoplus I(i,j,k) \quad (5)$$

$$I(i,j,k) = I(i,j,k) \quad (6)$$

5. Experiments and results

In this section, we present the result of six new experiments. In **Table 1**, we report the references and the size of the HI.

Number	Name and reference	Size
1	San Francisco[1]	702 × 1000 × 148
2	Urban [35, 36]	307 × 307 × 210
3	Indian pine [37]	145 × 145 × 220
4	Jasper Ridge [35, 36]	100 × 100 × 224
5	University of Pavia[2]	610 × 340 × 103
6	Samson [35, 36]	95 × 95 × 156

[1]*From Ref. [38].*
[2]*From Ref. [39].*

Table 1.
HI data set.

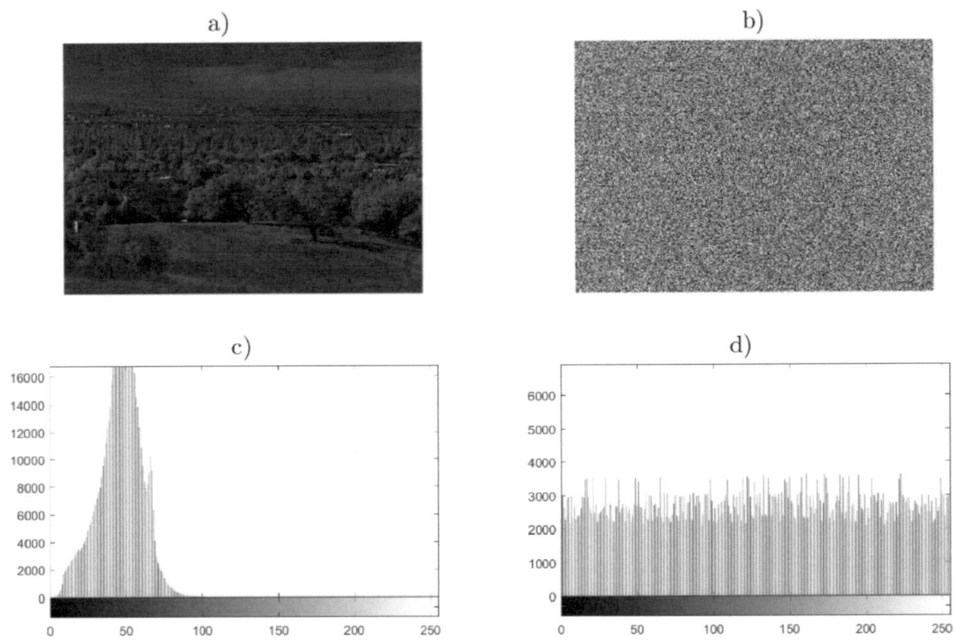

Figure 10.
Experiment 1: image comparison. a) Original image. b) Encrypted image. c) Original histogram. d) Encrypted histogram.

In the first experiment, we present the results of encrypting the "San Francisco" image. In **Figure 10**, we present a comparison of the ninety third channel of the original image and the cipher one. The image was loaded to the GPU in 0.4439 s and encrypted in 0.0094 s and then downloaded to the CPU memory in 0.21585 s.

The original image has an entropy of 6.1057; after the encryption, we get an entropy of 7.9985. In **Figure 11**, we show a comparison of the hyper-histograms of the original and encrypted image.

In the second experiment, we present the results of encrypting the "Urban" image. In **Figure 12**, we present a comparison of the sixtieth channel of the original image and the cipher one. The image was loaded to the GPU in 0.0118 s and encrypted in 0.0023 s and then downloaded to the CPU memory in 0.0435 s.

The original image has an entropy of 6.7452; after the encryption, we get an entropy of 7.9987. In **Figure 13**, we show a comparison of the hyper-histograms of the original and encrypted image.

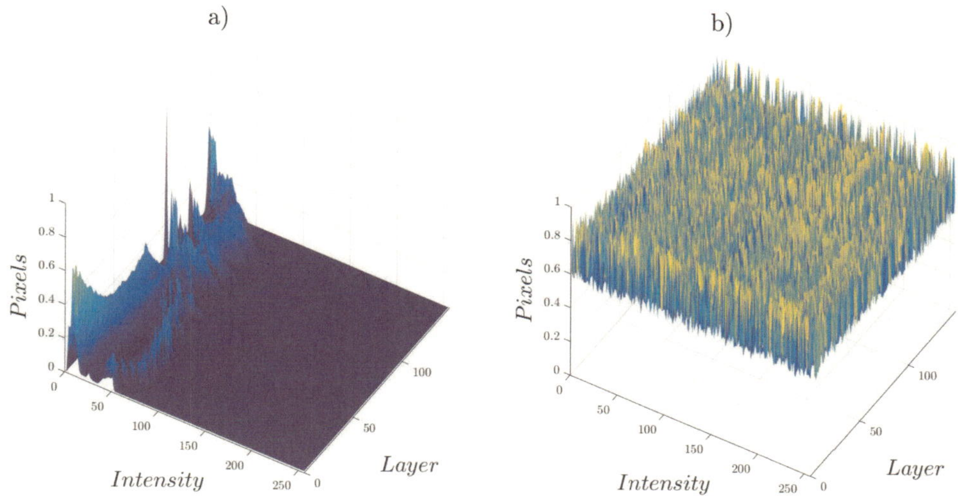

Figure 11.
Experiment 1: hyper-histogram comparison. a) Original Hyperhistogram. b) Encrypted Hyperhistogram.

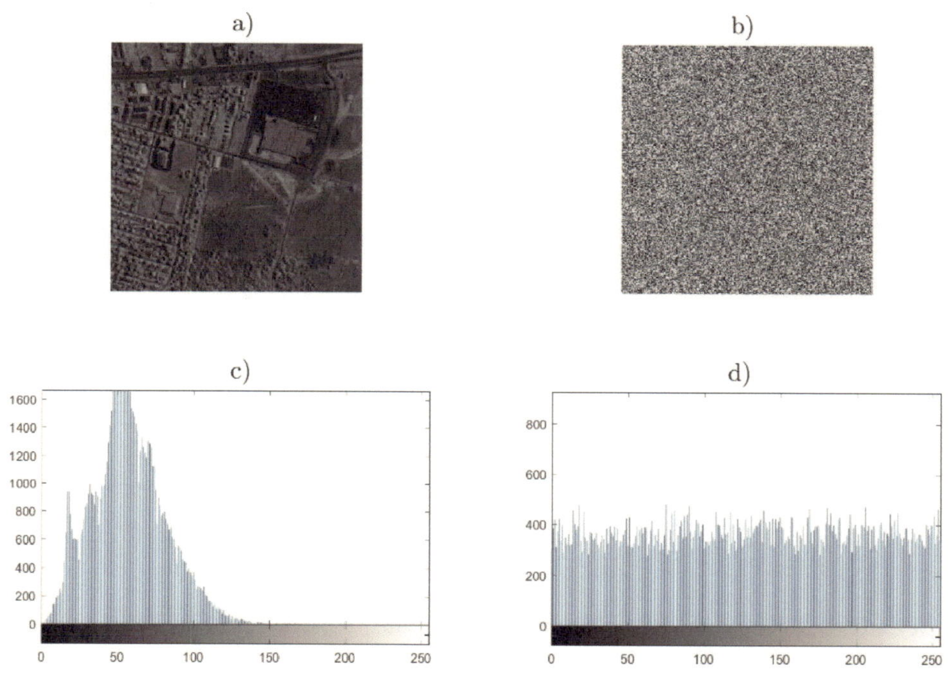

Figure 12.
Experiment 2: image comparison. a) Original image. b) Encrypted image. c) Original histogram. d) Encrypted histogram.

In experiment 3, we present the results of encrypting the "Indian pine" image. In **Figure 14**, we present a comparison of the forty-fifth channel of the original image and the cipher one. The image was loaded to the GPU in 0.0035 s and encrypted in 0.0010 s and then downloaded to the CPU memory in 0.0124 s.

The original image has an entropy of 6.3325; after the encryption, we get an entropy of 7.9985. In **Figure 15**, we show a comparison of the hyper-histograms of the original and encrypted image.

In experiment 4, we present the results of encrypting the "Jasper Ridge" image. In **Figure 16**, we present a comparison of the forty seventh channel of the original

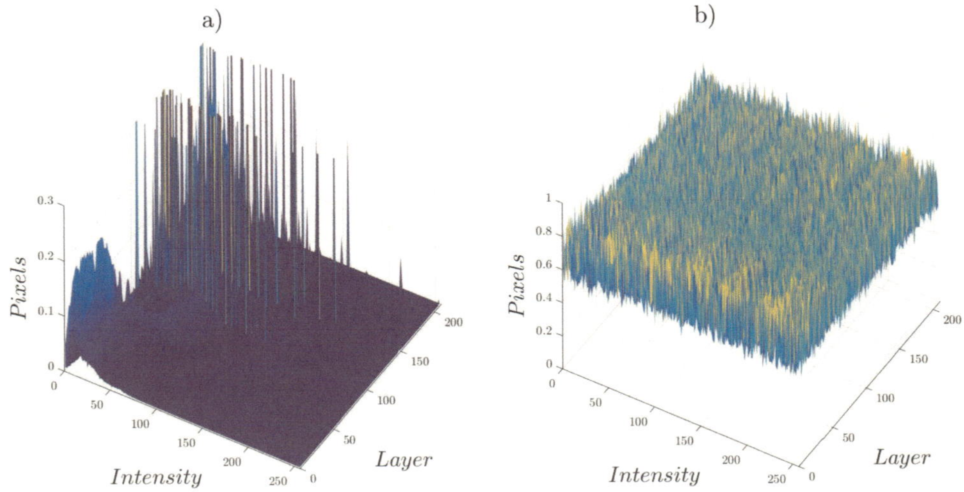

Figure 13.
Experiment 2: hyper-histogram comparison. a) Original Hyperhistogram. b) Encrypted Hyperhistogram.

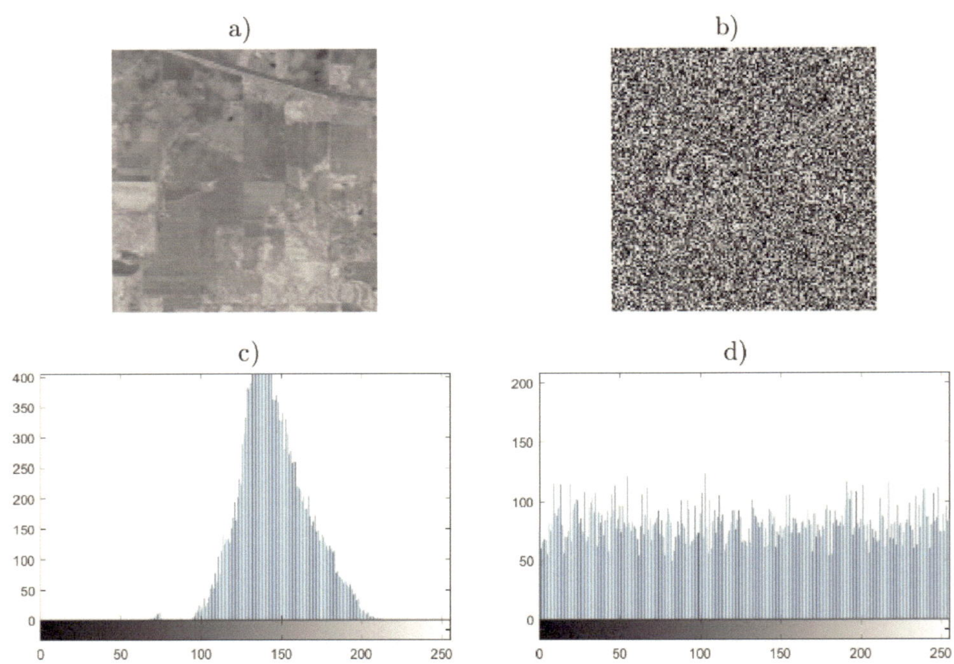

Figure 14.
Experiment 3: image comparison. a) Original image. b) Encrypted image. c) Original histogram. d) Encrypted histogram.

image and the cipher one. The image was loaded to the GPU in 0.0020 s and encrypted in 0.009 s and then downloaded to the CPU memory in 0.0055 s.

The original image has an entropy of 6.8918; after the encryption, we get an entropy of 7.9983. In **Figure 17**, we show a comparison of the hyper-histograms of the original and encrypted image.

In experiment 5, we present the results of encrypting the "University of Pavia" image. In **Figure 18**, we present a comparison of the 62nd channel of the original image and the cipher one. The image was loaded to the GPU in 0.0122 s and encrypted in 0.0026 s and then downloaded to the CPU memory in 0.0545 s.

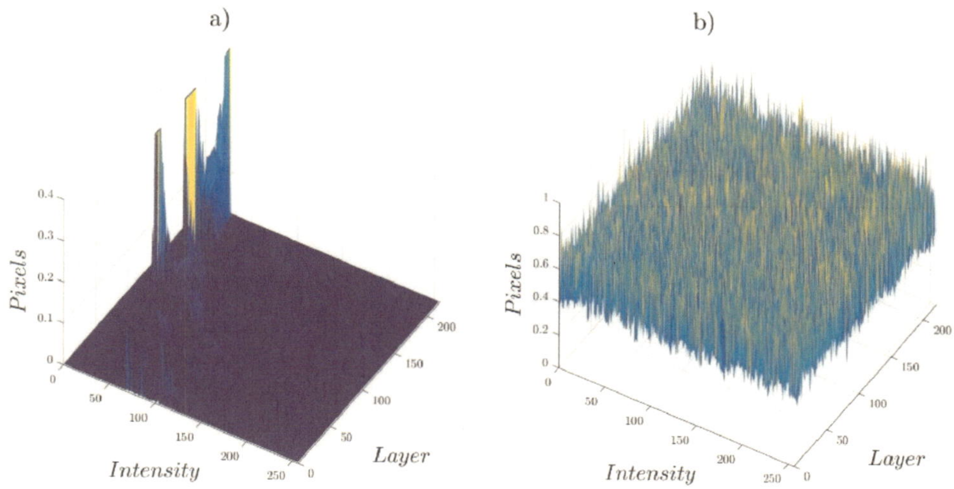

Figure 15.
Experiment 3: hyper-histogram comparison. a) Original Hyperhistogram. b) Encrypted Hyperhistogram.

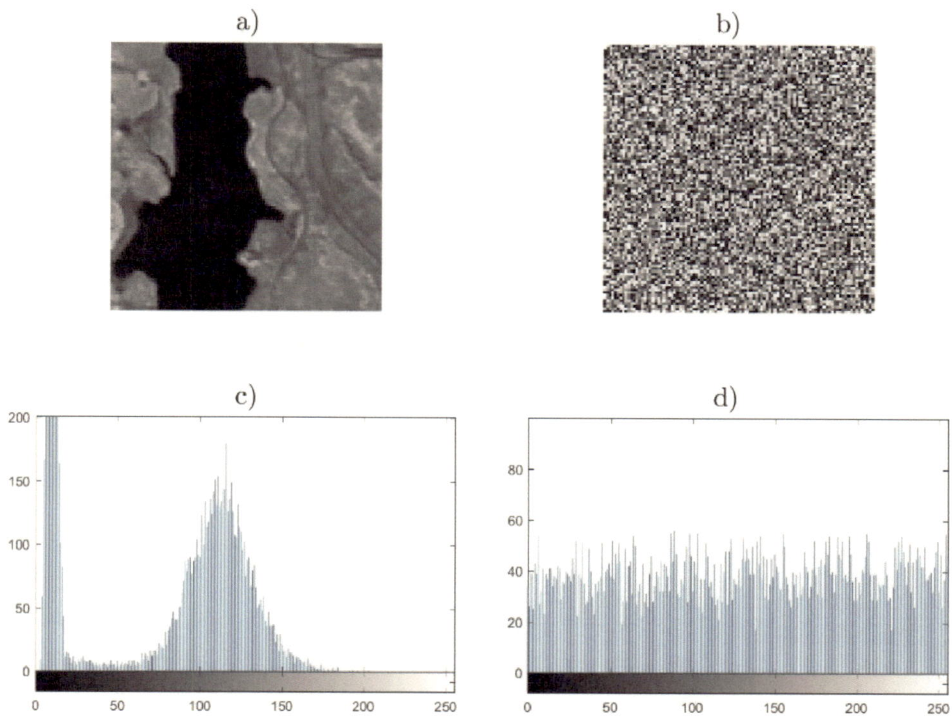

Figure 16.
Experiment 4: image comparison. a) Original image. b) Encrypted image. c) Original histogram. d) Encrypted histogram.

The original image has an entropy of 6.6827; after the encryption, we get an entropy of 7.9986. In **Figure 19**, we show a comparison of the hyper-histograms of the original and encrypted image.

In experiment 6, we present the results of encrypting the “Samson” image. In **Figure 20**, we present a comparison of the hundredth channel of the original image and the cipher one. The image was loaded to the GPU in 0.0013 s and encrypted in 0.0007 s and then downloaded to the CPU memory in 0.0036 s.

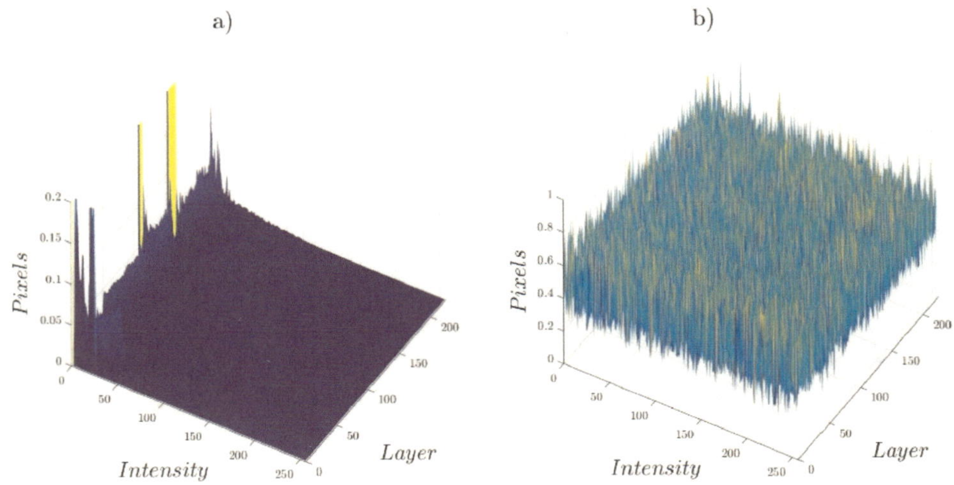

Figure 17.
Experiment 4: hyper-histogram comparison. a) Original Hyperhistogram. b) Encrypted Hyperhistogram.

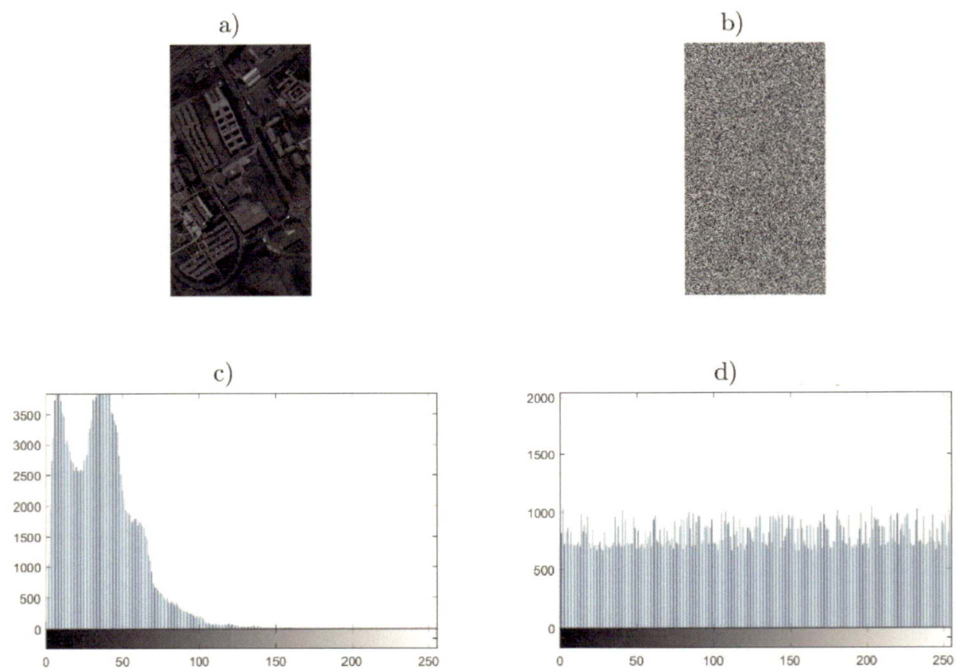

Figure 18.
Experiment 5: image comparison. a) Original image. b) Encrypted image. c) Original histogram. d) Encrypted histogram.

The original image has an entropy of 6.5833; after the encryption, we get an entropy of 7.9981. In **Figure 21**, we show a comparison of the hyper-histograms of the original and encrypted image.

In **Table 2**, we summarize the results of the encryption. Notice that the encryption process results in high entropy close to eight (the maximum for 2^8 representation). The entropy is a measure of uncertainty defined in Eq. (7), where s_i is a symbol or codification and $P(s_i)$ is the probability of s_i to appear. The number of bits for the representation, in this case eight, is depicted by b:

DOI: http://dx.doi.org/10.5772/intechopen.88980

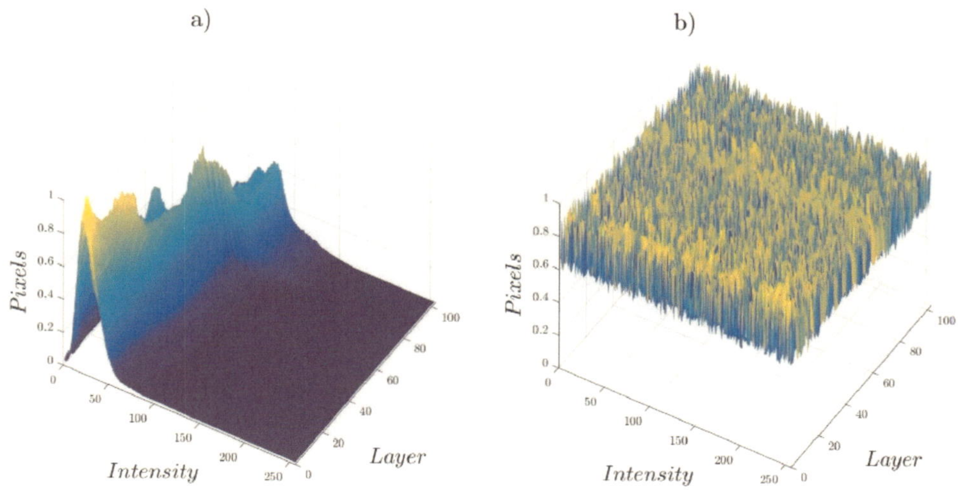

Figure 19.
Experiment 5: hyper-histogram comparison. a) Original Hyperhistogram. b) Encrypted Hyperhistogram.

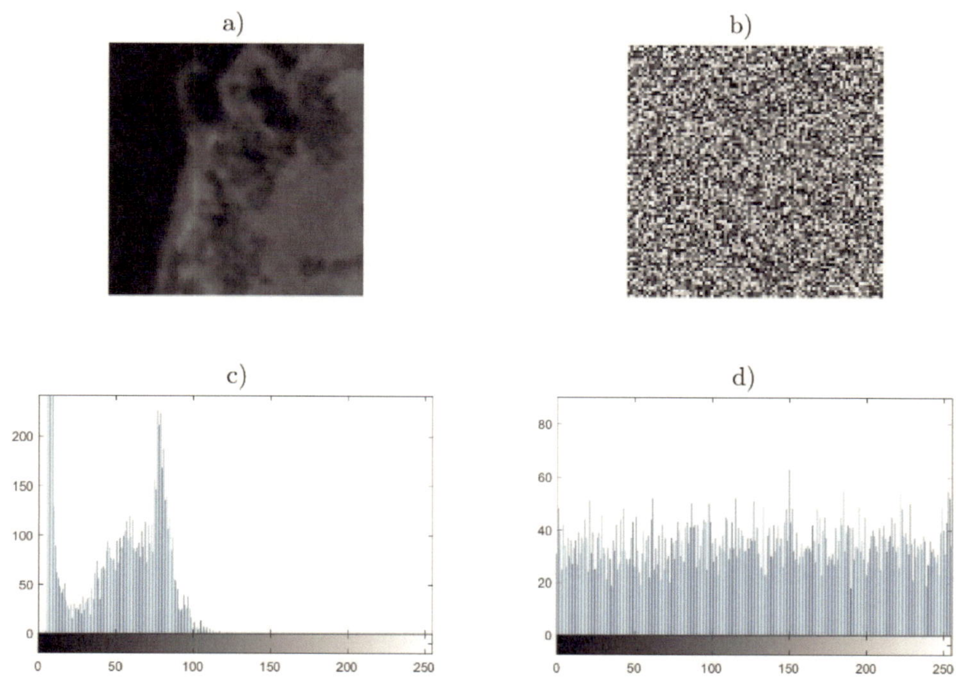

Figure 20.
Experiment 6: image comparison. a) Original image. b) Encrypted image. c) Original histogram. d) Encrypted histogram.

$$H(s) = -\sum_{i=0}^{2^b-1} P(s_i) \log_2(P(s_i)) \tag{7}$$

The proposed algorithm takes advantage of the multidimensionality nature of the HI. This leads to a high-performance algorithm that can encrypt images quickly. Notice that the load to GPU and download to CPU processes take most of the time. This could be avoided with a direct acquisition of the HI from the GPU, but this problem is beyond the scope of this chapter.

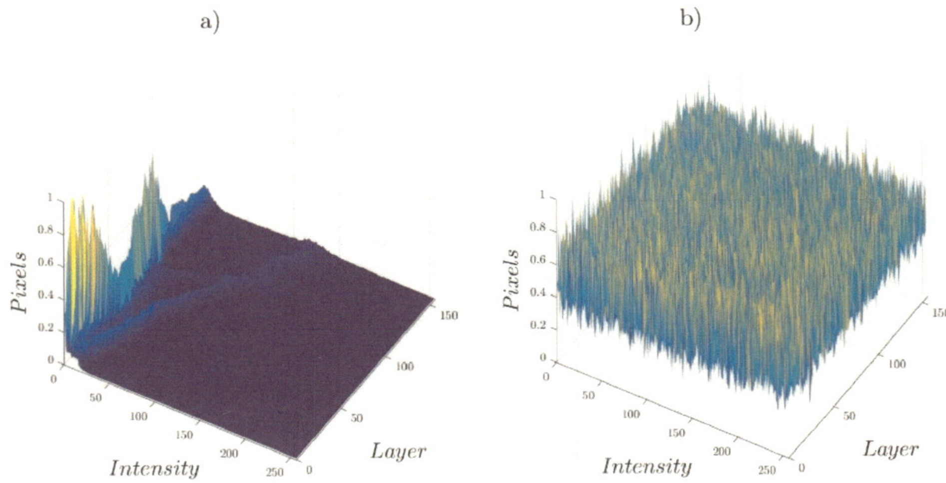

Figure 21.
Experiment 6: hyper-histogram comparison. a) Original Hyperhistogram. b) Encrypted Hyperhistogram.

Number	Name	Load (s)	Encrypt (s)	Download (s)	Total (s)	Original entropy	Encrypted entropy	Decrypt (s)
1	San Francisco	0.4439	0.0099	0.2158	0.6696	6.1057	7.9985	0.0095
2	Urban	0.0118	0.0023	0.0435	0.0576	6.7452	7.9987	0.0031
3	Indian pine	0.0035	0.0010	0.0124	0.0169	6.3325	7.9985	0.0012
4	Jasper Ridge	0.0020	0.0009	0.0055	0.0084	6.8918	7.9983	0.0009
5	University of Pavia	0.0122	0.0026	0.0545	0.0693	6.6827	7.9986	0.0023
6	Samson	0.0013	0.0007	0.0036	0.0056	6.5833	7.9981	0.0008

Table 2.
Encryption results.

6. Conclusions and future work

In this chapter, we have reviewed the proposed method in [8] with a full review of the source code. The proposed algorithm is a parallel encryption scheme specific for hyperspectral images. This scheme takes advantage of the multidimensionality nature of the HI, where every pixel of every channel is encrypted in parallel. These features lead to a high-performance algorithm capable of encrypting HI in a short time.

The algorithm uses four chaotic systems, the first three for generating a chaotic cube mask of the same size of the HI and the fourth one to generate a chaotic S-box to aggregate confusion in the ciphertext. As the experimentation has shown, the algorithm can generate high-entropy ciphertext. This algorithm is recommended for fast encryption in ciphertext-only attacks.

As future work, most of the computational time is wasted in the load to the memory of the HI image. A system to load directly to the GPU memory is desirable. Also, we recommend the readers to explore new ways of combining different chaotic signals for high-dimensional masks. Furthermore, an extension of this algorithm for working with a compatible compression algorithm is desired.

Author details

Carlos Villaseñor, Javier Gomez-Avila, Nancy Arana-Daniel*, Alma Y. Alanis and Carlos Lopez-Franco
Centro Universitario de Ciencias Exactas e Ingenierías, Universidad de Guadalajara, Guadalajara, Mexico

*Address all correspondence to: nancyaranad@gmail.com

References

[1] Shippert P. Why use hyperspectral imagery? Photogrammetric Engineering and Remote Sensing. 2004;**70**(4):377-396

[2] Paar C, Pelzl J. Understanding Cryptography: A Textbook for Students and Practitioners. Berlin/Heidelberg, Germany: Springer Science & Business Media; 2009

[3] Biryukov A, Khovratovich D. Related-key cryptanalysis of the full AES-192 and AES-256. In: International Conference on the Theory and Application of Cryptology and Information Security. New York, USA: Springer; 2009. pp. 1-18

[4] Thulasimani L, Madheswaran M. A single chip design and implementation of aes-128/192/256 encryption algorithms. International Journal of Engineering, Science and Technology. 2010;**2**(5):1052-1059

[5] Moh'd A, Jararweh Y, Tawalbeh L. AES-512: 512-bit advanced encryption standard algorithm design and evaluation. In: 2011 7th International Conference on Information Assurance and Security (IAS). 2011

[6] Somani U, Lakhani K, Mundra M. Implementing digital signature with RSA encryption algorithm to enhance the data security of cloud in cloud computing. In: 2010 First International Conference On Parallel, Distributed and Grid Computing (PDGC 2010). 2010

[7] Kocarev L. Chaos-based cryptography: A brief overview. IEEE Circuits and Systems Magazine. 2001;**1**(3):6-21

[8] Villaseñor C, Gutierrez-Frias EF, Arana-Daniel N, Alanis AY, Lopez-Franco C. Parallel crossed chaotic encryption for hyperspectral images. Applied Sciences. 2018;**8**:1-12

[9] Grahn H, Geladi P. Techniques and Applications of Hyperspectral Image Analysis. Hoboken, New Jersey, USA: John Wiley & Sons; 2007

[10] Alam MS, Sidike P. Trends in oil spill detection via hyperspectral imaging. 2012 7th International Conference on Electrical and Computer Engineering. 2012

[11] Goetz AF, Vane G, Solomon JE, Rock BN. Imaging spectrometry for earth remote sensing. Science. 1985; **228**(4704):1147-1153

[12] Govender M, Chetty K, Bulcock H. A review of hyperspectral remote sensing and its application in vegetation and water resource studies. Water Research Commission (WRC). 2007;**33**(2):145-152

[13] Adam E, Mutanga O, Rugege D. Multispectral and hyperspectral remote sensing for identification and mapping of wetland vegetation: A review. Wetlands Ecology and Management. 2010;**18**(3):281-296

[14] Liang H. Advances in multispectral and hyperspectral imaging for archaeology and art conservation. Applied Physics A. 2012;**106**(2):309-323

[15] Fischer C, Kakoulli I. Multispectral and hyperspectral imaging technologies in conservation: Current research and potential applications. Studies in Conservation. 2006;**51**(1):3-16

[16] Padoan R, Steemers T, Klein M, Aalderink B, De Bruin G. Quantitative hyperspectral imaging of historical documents: Technique and applications. In: Art Proceedings. 2008. pp. 25-30

[17] Clark RN, Swayze GA. Mapping minerals, amorphous materials, environmental materials, vegetation, water, ice and snow, and other materials: The USGS Tricorder algorithm. 1995

[18] Carrasco O, Gomez RB, Chainani A, Roper WE. Hyperspectral imaging

applied to medical diagnoses and food safety. In: Geo-Spatial and Temporal Image and Data Exploitation III. 2003

[19] Lu G, Fei B. Medical hyperspectral imaging: A review. Journal of Biomedical Optics. 2014;**19**(1):1-24

[20] Afromowitz MA, Callis JB, Heimbach DM, DeSoto LA, Norton MK. Multispectral imaging of burn wounds: A new clinical instrument for evaluating burn depth. IEEE Transactions on Biomedical Engineering. 1988;**35**(10): 842-850

[21] Gowen AA, O'Donnell CP, Cullen PJ, Downey G, Frias JM. Hyperspectral imaging—An emerging process analytical tool for food quality and safety control. Trends in Food Science and Technology. 2007;**18**(12): 590-598

[22] Feng Y-Z, Sun D-W. Application of hyperspectral imaging in food safety inspection and control: A review. Critical Reviews in Food Science and Nutrition. 2012;**52**(11):1039-1058

[23] Kuula J, Pölönen I, Puupponen H-H, Selander T, Reinikainen T, Kalenius T, et al. Using VIS/NIR and IR spectral cameras for detecting and separating crime scene details. In: Sensors, and Command, Control, Communications, and Intelligence (C3I) Technologies for Homeland Security and Homeland Defense XI, International Society for Optics and Photonics. 2012

[24] Edelman G, Gaston E, Van Leeuwen T, Cullen P, Aalders M. Hyperspectral imaging for non-contact analysis of forensic traces. Forensic Science International. 2012; **233**(3):28-39

[25] Obregón-Pulido G, Gasca A, Solis-Perales G. Secure communications using different type of chaotic systems. Multimedia, Communication and Computing Application. 2015;**1**(2):1-13

[26] Pecora LM, Carroll TL. Synchronization in chaotic systems. Physical Review Letters. 1990; **64**(1990):8-21

[27] Zhang L, Zhang Y. Research on lorenz chaotic stream cipher. In: Proceedings of 2005 IEEE International Workshop on VLSI Design and Video Technology 2005. 2005

[28] Jia X. Image encryption using the Ikeda map. In: 2010 International Conference on Intelligent Computing and Cognitive Informatics. 2010

[29] Sathishkumar G, Bhoopathy K, Sriraam N. Image encryption based on diffusion and multiple chaotic maps. International Journal of Network Security & Its Applications. 2011;**3**(2): 181-194

[30] Sankpal PR, Vijaya P. Image encryption using chaotic maps: A survey. In: 2014 Fifth International Conference on Signal and Image Processing; IEEE. 2014. pp. 102-107

[31] Su Z, Lian S, Zhang G, Jiang J. Chaos-based video encryption. In: Chaos-Based Cryptography. New York, USA: Springer; 2011. pp. 205-226

[32] Shang F, Sun K, Cai Y. An efficient MPEG video encryption scheme based on chaotic cipher. In: 2008 Congress on Image and Signal Processing; IEEE. 2008. pp. 12-16

[33] Lian S, Sun J, Wang Z, Dai Y. A fast video encryption scheme based-on chaos. In: ICARCV 2004 8th Control, Automation, Robotics and Vision Conference; IEEE. 2004. pp. 126-131

[34] Mroczkowski P. Generating pseudorandom S-boxes—A method of improving the security of cryptosystems based on block ciphers. Journal of Telecommunications and Information Technology. 2009;**2**:74-79

[35] Zhu F, Wang Y, Xiang S, Fan B, Pan C. Structured sparse method for hyperspectral unmixing. ISPRS Journal of Photogrammetry and Remote Sensing. 2014;**88**:101-118

[36] Zhu F, Wang Y, Xiang S, Fan B, Pan C. Spectral unmixing via data-guided sparsity. IEEE Transactions on Image Processing. 2014:5412-5427

[37] Baumgardner MF, Biehl LL, Landgrebe DA. 220 Band AVIRIS Hyperspectral Image Data Set: June 12, 1992 Indian Pine Test Site 3. [Online]. 2015. Available from: https://purr.purdue.edu/publications/1947/1 [Accessed: 10 June 2019]

[38] Database 4 of imageVal Consulting. Available from: http://www.imageval.com

[39] Hyperspectral Remote Sensing Scenes. Available from: http://www.ehu.eus/ccwintco/index.php?title=Hyperspectral_Remote_Sensing_Scenes

Section 2

Applications of Hyperspectral Image Processing

Chapter 5

NIR Hyperspectral Imaging for Mapping of Moisture Content Distribution in Tea Buds during Dehydration

Keqiang Yu, Yanru Zhao, Xiaoli Li and Yong He

Abstract

This work employed hyperspectral imaging technique to map the spatial distribution of moisture content (MC) in tea buds during dehydration. Hyperspectral images (874–1734 nm) of tea buds were acquired in six dehydrated periods (0, 3, 6, 9, 14 and 21 min) at 80°C. The spectral reflectance of tea buds were extracted from region of interests (ROIs) in the hyperspectral images. Competitive adaptive reweighted sampling (CARS) was used to select effective wavelengths (EWs) and ten representing the wavelengths were selected. The quantitative relationship between spectral reflectance and the measured MC values of tea buds was built using partial least square regression (PLSR) based on full spectra and EWs. The quantitative model established using EWs, which had a result of coefficient of correlation (R_P) of 0.941 and root mean square error of prediction (RMSEP) of 5.31%, was considered as the optimal model for mapping MC distribution. The optimal model was finally applied to predict the MC of each pixel within of the tea bud sample and built the MC distribution maps by utilization of a developed image processing procedure. Results demonstrated that the hyperspectral imaging technique has the potential of mapping the MC spatial distribution in tea buds in dehydrated process.

Keywords: NIR hyperspectral imaging, tea buds, moisture content, spatial distribution, dehydration process

1. Introduction

Tea, one of the most popular beverages worldwide, is of great interest due to its beneficial medicinal properties [1, 2]. Tea products are mainly made from the processed tea buds or fresh tea leaves of a plant called *Camellia sinensis*. In the tea processing, a great number of moisture are always changed along with a series of physical and chemical reactions. Especially in the drying stage with thermochemical reactions under high temperature, variations of moisture content (MC) in tea can directly affect smell, taste and others quality characteristics. With the growing consumption of the tea products, high qualities of tea products become more and more important nowadays. Therefore, in order to produce the high quality of tea products and prolong its shelf life, the determination of MC distribution in tea is quite meaningful in modern society.

The conventional way to analyze MC includes the gravimetric method, oven-dehydrated [3], freeze-dehydrated or lyophilization [4], electronic moisture analyzer [2], and so on. Those methods are time-consuming, tedious and fail to meet the requirements of real-time, on-line detection of MC in tea processing. In addition, the same sample cannot be reused for any other purpose and those methods may debase the quality of tea products through directly touching way.

In recent years, spectroscopy technique has proved to be a powerful tool for detecting the MC in tea products and agricultural sideline products. For example, Mizukami et al. [5] developed a new method for measuring the moisture in tea leaves using an electrical spectroscopy. Diffuse reflectance spectroscopy combined with chemometric analysis were employed to investigate MCs in tea [6]. Sinija and Mishra [2] employed Fourier transform near infrared (FTNIR) spectroscopy to measure MC in green tea. However, spectroscopy technique is not able to provide spatial information of quality parameters, which greatly limited its application to quantify spatial distribution.

Hyperspectral imaging, a powerful analytical tool, has attracted a great deal of attention for the safety detection of agricultural and sideline products. It integrated conventional spectral information and digital imaging into one system, which made it possible for providing both spectral and spatial information of an object simultaneously [7]. Over the past several years, hyperspectral imaging has many potential applications for quantifying and controlling of quality parameters with good precision. It is widely applying in evaluation of various agricultural products, such as beef [8], pork [9] and lamb [10], moisture in prawn [11], mushroom [12], moisture in banana [13], strawberry [14] and maturity and firmness of apple [15], and texture analysis to classify green tea [16].

However, to the best of our knowledge, applying the hyperspectral imaging technique to determine the moisture distribution in tea buds has not been found to date. There are also some broadband peaks occurring in the NIR region related to the overtone and combination vibrations of hydrogen containing bonds, such as O–H, C–H, and N–H [11]. The presence of water (O-H) in the tea buds showed two feature wavelengths at 980 and 1450 nm (O–H stretching second and first overtones) in NIR region. So, this research employed the NIR hyperspectral imaging for predicting and mapping the distribution of MC in tea buds. The steps of the work are to: (1) obtain hyperspectral image of tea buds in NIR region of 874–1734 nm and measure the MCs of tea bud samples in dehydrated process; (2) extract spectral data of the region of interests (ROIs) from the acquired hyperspectral images; (3) select the effective wavelengths which carried the most valuable information related to MC prediction and build the quantitative models; (4) develop an image processing procedure for mapping the spatial distribution of MC in tea buds. The main steps involved in building MC distribution maps are presented in **Figure 1**.

2. Materials and methods

2.1 Pretreatment of tea buds samples

In this research, buds of tea bushes (*C. sinensis* cv. Longjing 43) were prepared for the experiment. Tea bushes were planted about 6 years in the Zijingang campus of Zhejiang University, Hangzhou (30°16′N, 120°20′E), China. Each tea bud sample contained three fresh leaf blades. A total of 216 tea bud samples were randomly collected from different tea clusters on 2 April 2013. At first, random 36 tea buds were selected for acquiring the hyperspectral images. Then the remaining 180 samples were randomly divided into five groups. These five groups of tea buds

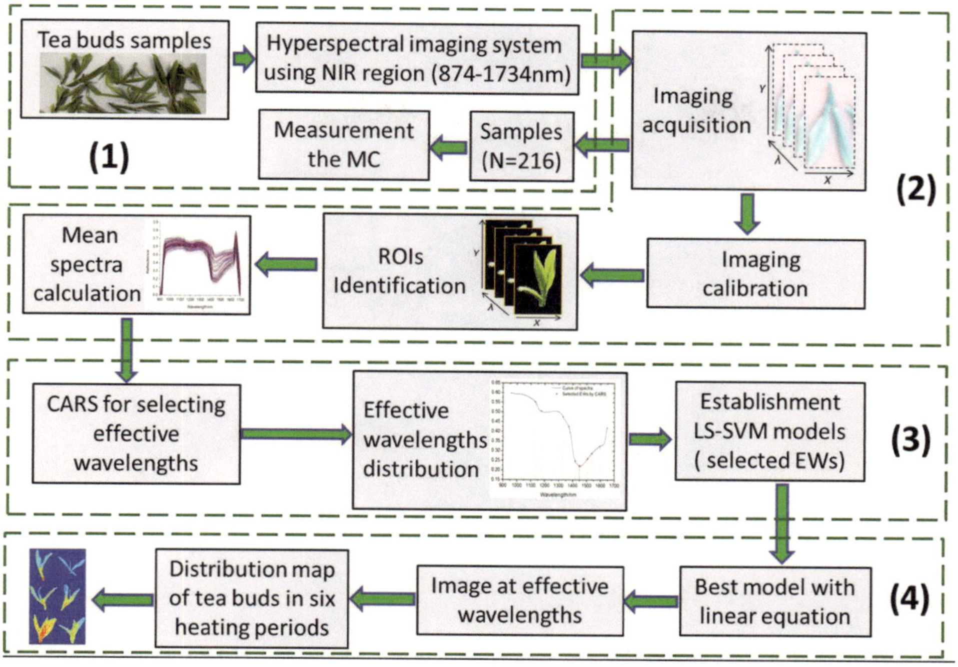

Figure 1.
Main steps for building of MC distribution maps in tea buds by using hyperspectral imaging, (1) pre-treatment of tea buds; (2) hyperspectral data pre-processing; (3) analyses of spectral data; (4) image post-processing for building the MC distribution maps of tea buds.

were implemented to dehydrate at 80°C by an attemperator (IKA@C-MAG HS4, Germany) for corresponding five dehydrated times (3, 6, 9, 14 and 21 min), respectively. Meanwhile, all samples were scanned by to acquire hyperspectral images in the corresponding dehydrated time.

After acquirement of hyperspectral images, the MC of all samples was measured by the gravimetric method according to the Chinese National Standard GB8304-87. In detail, all samples were dried in a constant temperature oven at 103°C for 18 h. Meanwhile, an electronic balance with accuracy of 0.0001 g was employed to weight all samples after acquiring hyperspectral images and drying. All the measurements were carried out in a room at approximate constant temperature of 25°C and relative humidity of 35–45%. In addition, all the 216 tea bud samples were divided into a calibration set (162 samples) and a prediction set (54 samples) by Kennard-Stone (K-S) algorithm [17].

2.2 Hyperspectral imaging acquiring equipment

In this study, a laboratory pushbroom hyperspectral imaging equipment (**Figure 2**) with reflectance mode was employed to scan all the samples. As Yu et al. [18] described, the core sensing components of the equipment consisted of several parts: a conveyor belt operated by a stepper motor (IRCP0076, Isuzu Optics Crop, Taiwan, China); an illumination unit assembled by two 150-W quartz tungsten halogen lamps (Fiber-Lite DC950 Illuminator, Dolan Jenner Industries Inc., USA); an imaging spectrograph (ImSpector N17E, Spectral Imaging Ltd., Finland) covering a spectral range of 874–1734 nm; a CCD camera (C8484-05, Hamamatsu, Hamamatsu city, Japan) coupled with a camera lens (OLES23; Specim, Spectral Imaging Ltd., Oulu, Finland) and a computer with the spectral-cube data acquisition software (Isuzu Optics Corp, Taiwan, China), which could set and adjust the

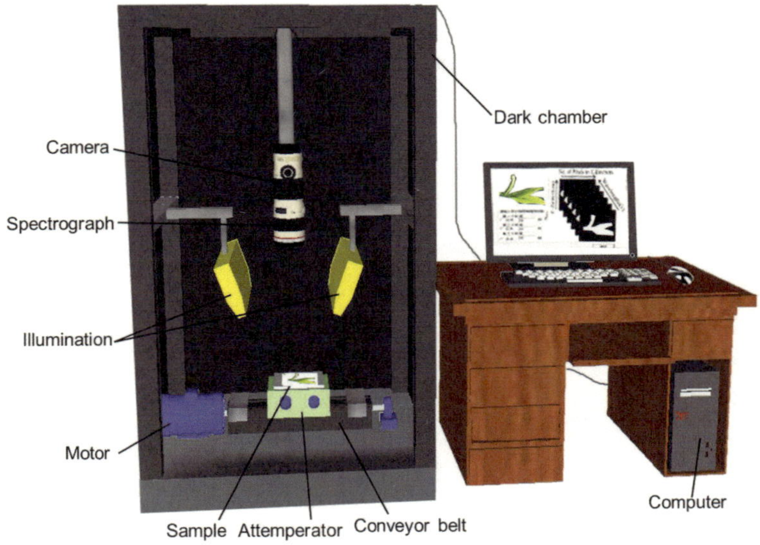

Figure 2.
Schematic diagram of the hyperspectral imaging equipment.

speed of conveyer belt, exposure time, binning mode, wavelength range, image acquisition, images calibration and so on. Overall, all the components (except computer) were fixed inside a dark chamber to avoid any stray light which might affect the veracity of hyperspectral imaging equipment.

In order to acquire clear and undistorted hyperspectral images, some parameters of the equipment needed to be adjusted before the images acquirement. Firstly, illumination unit should set an appropriate intensity and adjust a proper angel to make the light gather in a linear area of the conveyor belt just below the imaging spectrograph. Then, two reference reflectance panels with reflectance of 99.9 and 0% were adopted for dark and white reflectance calibration of sample. In this study, the distance between samples and the lens was 165 mm. All samples placed on the conveyor belt and moved at a speed of 14.5 mm/s to be scanned with an exposure time of 5 ms during the image acquisition. Gradually in line by line pattern, a hyperspectral image called "hypercube" with dimension of (x, y, λ) was built. In this study, the hyperspectral images were obtained with 320 pixels in x-direction, n-pixels in y-direction (based on the length of each sample) and 256 wavelengths in λ-direction.

2.3 Calibration of hyperspectral images

Because of the existence of dark current in CCD camera and the uneven intensity of illumination in different bands, several bands with weaker light intensity contained the bigger noises [19]. Based on this point, the raw hyperspectral images (I_{raw}) required to be calibrated and the calibration process could be finished using the following Eq. (1) [20, 21]:

$$R = \frac{I_{raw} - I_{dark}}{I_{white} - I_{dark}} \tag{1}$$

where, R were the calibrated hyperspectral images of the samples; I_{dark} were the dark reference images (~0% reflectance) obtained with light source off; I_{white} were the white reference images (~99% reflectance) acquired from a white reference ceramic tile. Then the calibrated images were used as the basis for subsequent processing and analysis.

2.4 ROIs identification and spectral data extraction

Spectral data were extracted by the region of interests (ROIs) function of ENVI software. An irregular ROI was identified by initially shape of tea bud in hyperspectral image. Then, the mean relative reflectance for each image by averaging the spectral responses of each pixel in the ROI was calculated. According to this procedure, a total of 216 mean reflectance spectra were obtained from the hyperspectral images of tea bud samples. Because of because the response of the CCD detector [8] and strong noise existence, the reflectance in two regions of 874–950 and 1670–1734 nm was rather low and littery. Therefore, hyperspectral images were resized to the spectral range of 950–1670 nm with a total of 214 wavebands.

2.5 Chemometric of spectral data processing

Competitive adaptive reweighted sampling (CARS), a novel algorithm for selecting important variables [22], was employed to select the effective wavelengths from the full range spectra of the calibration in this study. Details of the CARS methodology could be found in Li et al. [22].

Partial least square regression (PLSR), one of the most robust and reliable analytical tools for modeling, is a linear and supervised multivariate calibration method [23]. PLSR projects the spectral data onto a set of orthogonal factors called latent variables (LVs), and explores the optimal function by minimizing the error of sum squares (finding the optimal LVs), which is typically done by cross-validation [24]. The process of extracting the LVs should take the response variable into account. In this research, the quantitative model between the spectral reflectance and MCs was established using the PLSR.

The performance of a calibration model is usually evaluated according to coefficients of correlation (R) and root mean square error (RMSE) in calibration (R_C, RMSEC), in cross-validation (R_{CV}, RMSECV) and in prediction (R_P, RMSEP). Generally speaking, a model with larger values of R_C, R_{CV} and R_P, smaller values of RMSEC, RMSECV and RMSEP is wonderful, and it has a small difference between RMSEC, RWSECV and RMSEP.

In this research, data extractions, statistical calculations and multivariate data analyses were executed with ENVI 4.6 software (ITT Visual Information Solutions, Boulder, CO, USA), "The Unscrambler X 10.1" (CAMO PROCESS AS, Oslo, Norway) and MATLAB 7.8 (R2009a) software (The Math Works, Inc., Natick, MA, USA). The developed procedures for mapping MC distribution were completed in MATLAB.

3. Results and discussion

3.1 Spectral features of tea buds and statistics of measured MC

In general, NIR spectra region contained rich information relevant to hydrogen containing bonds than others spectra region [24]. To compare spectral trends over six dehydrated periods, the mean spectral values of the pixels within the ROI of tea bud samples were calculated. And those values exhibited some variances and overlays between two adjacent dehydrated periods (not given here). The mean spectral reflectance curves are illustrated in **Figure 3**. There were also some broadband

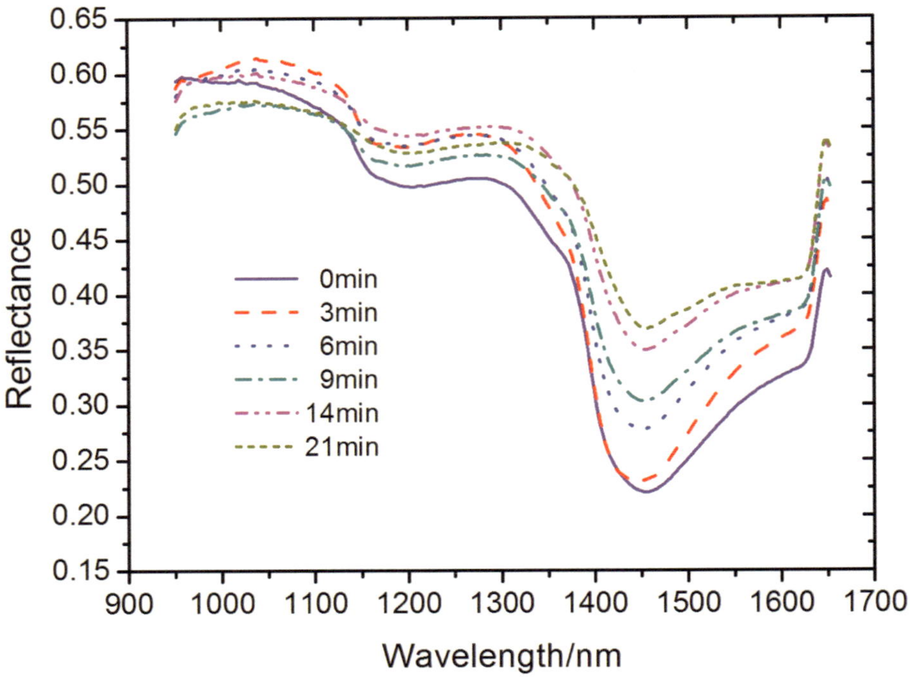

Figure 3.
Mean spectral reflectance of tea buds in six dehydrated periods.

peaks occurring in the NIR region related to the overtone and combination vibrations of hydrogen containing bonds, such as O-H, C-H, and N-H [25]. As is shown in **Figure 3**, the existence of water in the tea buds showed two feature wavelengths around 980 and 1450 nm (O-H stretching second and first overtones). Additionally, the absorption peak around 1200 nm (C-H stretching second overtone) was due to organic matter content in tea bud. Because of the complex chemical compositions (maybe including C-H and N-H) in tea buds, it is hard to find a clear trend of curves over MC within 950–1100 nm. However, it was worth noting that the spectral reflectance curves over MC showed a clear upward trend in the vicinity of 1450–1650 nm during the dehydrated processing in five periods (0, 3, 6, 9, 14 and 21 min).

In addition, **Figure 4** summarizes the statistics of MCs including mean, max, min and standard deviation (SD) values of tea bud samples in six dehydrated periods. It could be concluded that the mean, max and min values of MC appeared an obviously decreasing trend. Especially in mean values of those groups, a remarkable gradient (declining about 10%) was easily observed.

3.2 Variables selection

In this study, CARS was employed to select the effective variables. During the CARS process, some key variables were survived, while incompetent variables were sifted out. **Figure 5** demonstrated the process of variable selection by CARS.

Figure 5(a) illustrated that the number of sampled variables decreased fast at the first stage of EDF and then slowly at the second stage of EDF, which demonstrated "fast selection" and "refined selection". And in **Figure 5(b)**, it was clearly that along with the number of sampling runs increased, RMSECV values first reduced in sampling runs 1–4, and then fluctuated in a gentle way in the sampling runs 5–33, finally in sampling runs 34–50 increased fast. In this process, most of

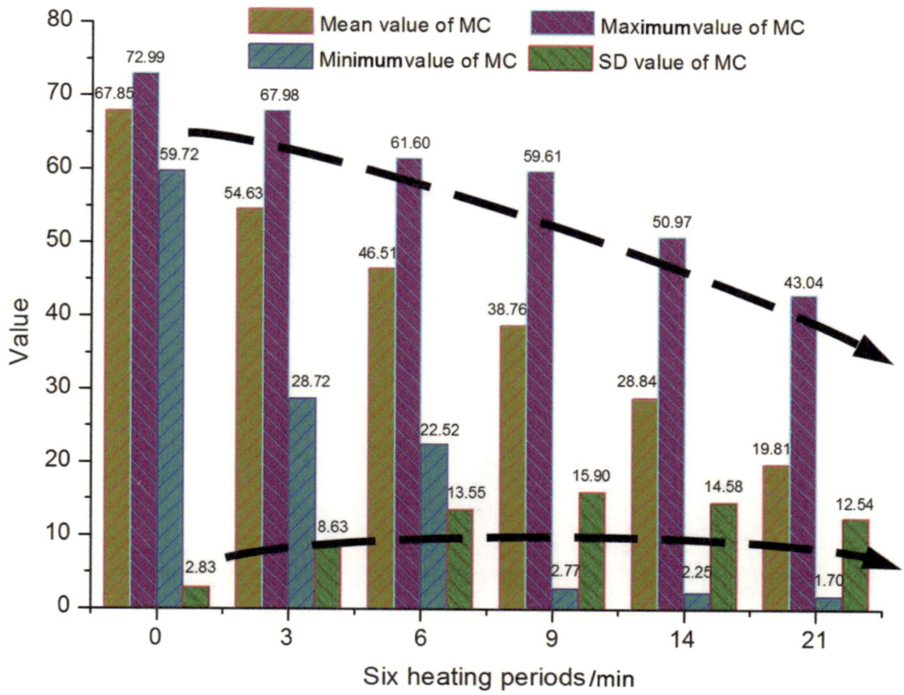

Figure 4.
Statistical results of measured MC of tea bud samples in six dehydrated periods.

uninformative variables were eliminated, and finally the RMSECV value increased because of the loss of some key variables [22]. The optimal variable subset was determined corresponding to the minimal 5-fold RMSECV value, and located by the vertical blue asterisk line in **Figure 5(c)**. Moreover, the regression coefficient path of each wavelength was also shown in **Figure 5(c)**. The variation of coefficient values of each variable was recorded by the colorful lines at different sampling runs. At the beginning of the each number sampling run, the absolute value of regression coefficient of each wavelength was very lowly. After that, values of some variables had a growing trend, while the rest of variables became smaller and smaller and turned into zero eventually (those were weeded out) because of their incompetence. In other words, the larger the absolute coefficient was, the more possibility the corresponding wavelength was able to survive.

Based on the calculation of CARS, ten wavelengths at 1133, 1173, 1332, 1372, 1419, 1446, 1450, 1507, 1538 and 1595 nm were identified as the EWs for predicting MC of tea buds. And the distribution of the selected EWs based on CARS was demonstrated in **Figure 6**.

Obviously, most of those selected EWs (1419, 1446, 1450 and 1507 nm) were scattered around the O-H stretching first overtones (1450 nm). Comparatively speaking, only two effective wavelengths (1133 and 1173 nm) were centered in C-H stretching second overtone (1200 nm), which might be related to organic matter of tea buds.

3.3 Modeling of MC in tea buds by PLSR

In this research, the multivariate models were established by PLSR algorithm with full spectra and EWs, respectively. In the PLSR model of calibration set, the quantitative relationship between the spectral reflectance and corresponding

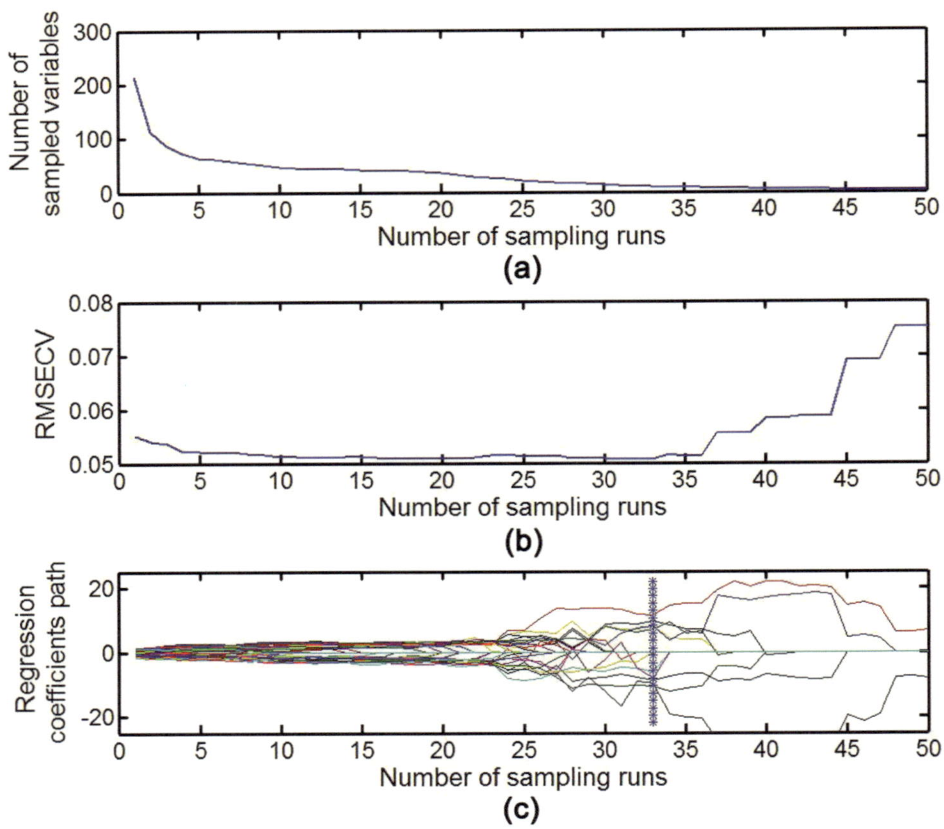

Figure 5.
Selection of effective wavelengths, (a) changing trend of the number of sampled variables; (b) 5-fold RMSECV values; (c) regression coefficients of each variable with the increasing of sampling runs, the line (marked by asterisk) denoted the optimal point where 5-fold RMSECV values achieved the lowest.

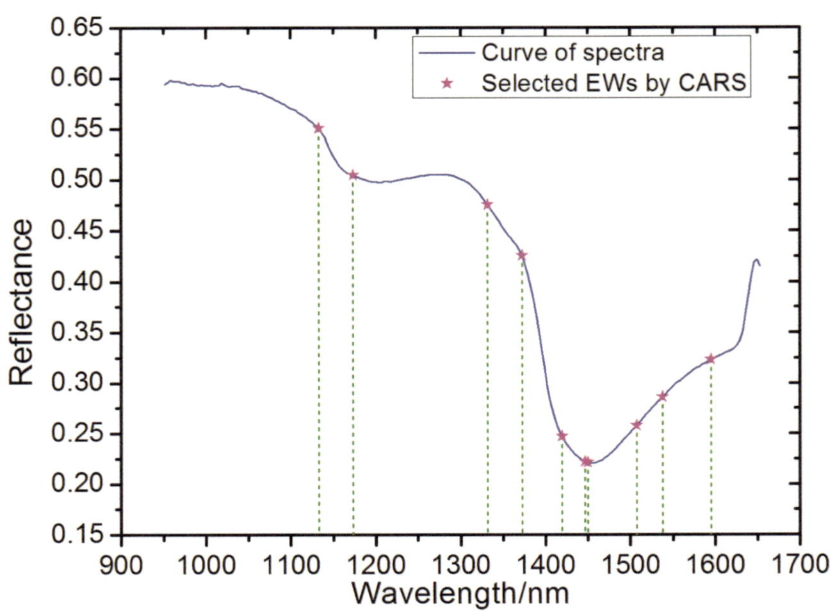

Figure 6.
The distribution of the selected effective wavelengths.

measured MC of tea bud samples was established. **Table 1** displayed the statistical results with respect to the prediction of MC in tea buds by using the full spectra and EWs.

Form **Table 1**, the PLSR model based on full spectra (F-PLSR) with $R_C = 0.956$, RMSEC = 5.22%, $R_{CV} = 0.908$, RMSECV = 5.93%, $R_P = 0.946$, RMSEP = 5.07% had the better result for predicting MC in tea buds. Compared to the F-PLSR model, the results of CARS-PLSR model had a slight drop in R_C, R_p of 0.008, 0.005, respectively. The above results indicated that models using full spectra for predicting MC of tea buds had excellent predictive accuracy and robustness.

Regretfully, full spectra had a high-dimensional data and F-PLSR model could not provide a simple linear function about the reflectance of spectral reflectance and MC of tea buds. On the contrary, the selected ten EWs had minimal redundancy and offered a commendable prediction performance ($R_P = 0.941$). Hence, CARS-PLSR model was considered as the ideal model for the predicting MC of tea buds during the six dehydrated periods. The obtained function Eq. (2) according to CARS-PLSR model was shown as follows:

$$\begin{aligned} Y_{mositure} = {} & 0.7649 + 7.8540\lambda_{1133nm} - 3.2826\lambda_{1173nm} \\ & - 13.9586\lambda_{1332nm} + 7.4947_{\lambda 1372nm} - 2.3084\lambda_{1419nm} \\ & + 3.1167\lambda_{1446nm} + 1.8836\lambda_{1450nm} - 5.9708\lambda_{1507nm} \\ & - 5.3227\lambda_{1538nm} + 19.9262\lambda_{1595nm} \end{aligned} \tag{2}$$

where, $\lambda_{i\ nm}$ was the spectral reflectance at the wavelength of i nm, and $Y_{mositure}$ was the predicted moisture content of tea buds. In addition, the obtained function was also taken for further analysis of mapping the spatial distribution of MC in tea buds.

3.4 Distribution maps of MC in tea buds

For predicting MC in all spots of the sample, the CARS-PLSR model was then transferred to each pixel of the image. After multiplying the model's regression coefficients by the spectrum of each pixel in the image [25], a prediction image (called distribution map) was built and exhibited the spatial distribution of MC of the sample. In the final distribution map, the pixels with similar spectral characteristics would generate the same predicted values of MC, which were led to a similar color in the acquired image [7, 25].

Figure 7 shows examples of spatial distribution maps of tea buds with different MC levels in six dehydrated periods. **Figure 7(a)** showed the pseudo-color images of six tea bud samples with different MC values. The values at the top of tea buds represented the average concentration of moisture in the whole samples. As seen

Models	Variable number	LVs	Calibration		Cross-validation		Prediction	
			R_C	RMSEC (%)	R_{CV}	RMSECV (%)	R_P	RMSEP (%)
F-PLSR	214	7	0.956	5.22	0.908	5.93	0.946	5.07
CARS-PLSR	10	5	0.948	5.15	0.921	5.43	0.941	5.31

Note: F-PLSR stood for meant the PLSR model established using the full spectra; CARS-PLSR represented the PLSR model built based on EWs selected by CARS.

Table 1.
The results of PLSR models for predicting MC in tea buds based on full spectra and the effective wavelengths.

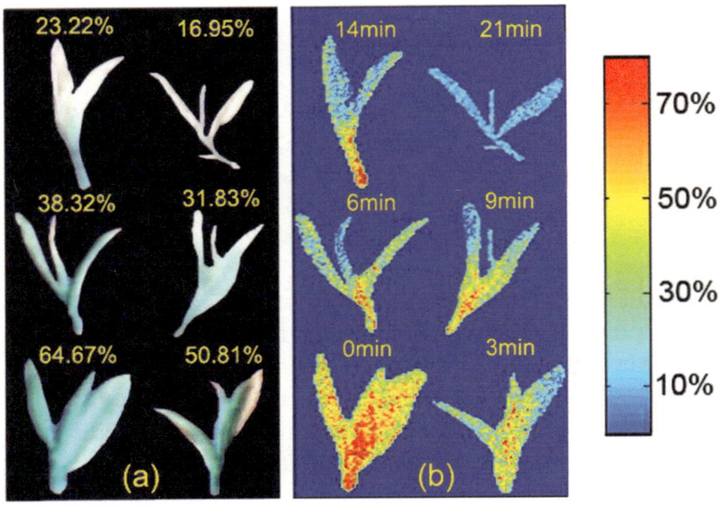

Figure 7.
Distribution maps of MC in tea buds in six dehydrated periods. (a) Pseudo-color image of the tea buds in three monochromatic images at 1399, 1197 and 995 nm; and (b) the spatial distribution maps of MC in six dehydrated periods.

clearly that the color of samples changed from emerald to silver along with MC values decrease. In addition, it was so hard to find out the difference in MC from point to point by naked eye from the pseudo-color image. Surprisingly, the different MC level among the samples was very obvious to be discerned from the final distribution maps as shown in **Figure 7(b)**. A linear color scale was generated with the different MC values from small to large shown in different color from blue to red. The MC of tea buds from high to low was displayed in different colors from red to blue. Meanwhile, the difference of MC level within a sample could be easily identified. A comprehensive map with different colors indicated that there were mixed components and heterogeneous distribution of MC in samples. Many pixels in fresh tea bud (0 min dehydrated) were red or orange because it had an average MC value of 64.67%. Along with the dehydrated time increased, the color of pixels changed from red, orange, pale bluish green to blue, which indicated the moisture of tea bud gradually lost. It was worth noticing that the edge of blade was turned blue firstly, then the vein, and finally the petiole. Especially in two dehydrated periods (9 and 14 min), the color of leaf petiole and blade was obviously different. This is a clear indication of MC status during the dehydrated process of tea buds.

In fact, the water in the blade is free water in mesophyll cells while it exists in a form of the bound water in xylem, leaf vein and petiole. Meanwhile, the free water can be easily dried in a short time at a lower temperature, in contrast evaporation of the bound water requires a long time at a higher temperature. So the MC of the blade is lower than that of leaf vein and petiole in the early stage of drying. Through visualization analysis, and spatial variation of MC can be intuitively detected which will provide vital information for understanding the drying dynamics of tea leaf and optimization of tea leaf drying process.

4. Conclusions

This study was conducted to evaluate the dominant position of hyperspectral imaging technique in NIR region for mapping spatial distribution of MC in tea buds during dehydration. The results demonstrated that as a promising technology, hyperspectral imaging could achieve the objective of mapping the MC distribution in tea bud during the drying process. In this research, the chemometric method of CARS was employed to select EWs. After that, PLSR algorithm was used to establish the quantitative relationship between the spectral reflectance and measured MC of tea buds. At last stage, the MC of all pixels in tea buds were calculated based on the optimal PLSR model. Meanwhile, the spatial distribution maps were built using a developed image processing procedure. The spatial variation of MC could expose the different MC within tea buds in different dehydrated periods, which applied an approach to kinetic analysis of MC in drying process, and provide important information for optimization of tea processing technic.

In further research, tea products with more types of sample and different geographical locations, ages and times should be taken into account to establish more robust and generate MC determination model, which could give more help for optimization of dehydrated process of agricultural products.

Acknowledgements

This work was supported by the National Natural Science Foundation of China (Program No: 61705188), Natural Science Basic Research Plan in Shaanxi Province of China (Program No. 2017JQ3008), China Postdoctoral Science Foundation (2017 M613218), Shaanxi Postdoctoral Science Foundation, the Fundamental Research Funds for the Central Universities (2452017125), the Doctoral Scientific Research Foundation of Northwest A&F University (2452016157), and the Key Laboratory of Agricultural Internet of Things, Ministry of Agriculture and Rural Affairs, P. R. China.

Conflict of interest

The authors have declared no conflict of interest.

Author details

Keqiang Yu[1,2,3*], Yanru Zhao[1,2,3], Xiaoli Li[4] and Yong He[4]

1 College of Mechanical and Electronic Engineering, Northwest A&F University, Yangling, Shaanxi, PR China

2 Key Laboratory of Agricultural Internet of Things, Ministry of Agriculture and Rural Affairs, Yangling, Shaanxi, PR China

3 Shaanxi Key Laboratory of Agricultural Information Perception and Intelligent Service, Yangling, Shaanxi, PR China

4 College of Biosystems Engineering and Food Science, Zhejiang University, Hangzhou, PR China

*Address all correspondence to: keqiang_yu@nwafu.edu.cn

References

[1] Yang CS, Maliakal P, Meng X. Inhibition of carcinogenesis by tea. Annual Review of Pharmacology and Toxicology. 2002;**42**(1):25-54. DOI: 10.1146/annurev.pharmtox.42.082101.154309

[2] Sinija VR, Mishra HN. FT-NIR spectroscopy for caffeine estimation in instant green tea powder and granules. LWT-Food Science and Technology. 2009;**42**(5):998-1002. DOI: 10.1016/j.lwt.2008.12.013

[3] Association of Official Analytical Chemists (AOAC). In: Horwitz W, editor. Official Methods of Analysis. Vol. 534. Washington, DC: AOAC; 1980. DOI: 10.1016/0165-9936(90)87098-7

[4] Greiff D. Protein structure and freeze-drying: The effects of residual moisture and gases. Cryobiology. 1971;**8**(2):145-152. DOI: 10.1016/0011-2240(71)90022-8

[5] Mizukami Y, Sawai Y, Yamaguchi Y. Moisture content measurement of tea leaves by electrical impedance and capacitance. Biosystems Engineering. 2006;**93**(3):293-299. DOI: 10.1016/j.biosystemseng.2005.12.009

[6] Li X, Xie C, He Y, Qiu Z, Zhang Y. Characterizing the moisture content of tea with diffuse reflectance spectroscopy using wavelet transform and multivariate analysis. Sensors. 2012;**12**(7):9847-9861. DOI: 10.3390/s120709847

[7] He H-J, Wu D, Sun D-W. Non-destructive and rapid analysis of moisture distribution in farmed Atlantic salmon (*Salmo salar*) fillets using visible and near-infrared hyperspectral imaging. Innovative Food Science and Emerging Technologies. 2013;**18**:237-245. DOI: 10.1016/j.ifset.2013.02.009

[8] Wu D, Wang S, Wang N, Nie P, He Y, Sun D-W, et al. Application of time series hyperspectral imaging (TS-HSI) for determining water distribution within beef and spectral kinetic analysis during dehydration. Food and Bioprocess Technology. 2012;**6**(11):2943-2958. DOI: 10.1007/s11947-012-0928-0

[9] Dissing BS, Papadopoulou OS, Tassou C, Ersbøll BK, Carstensen JM, Panagou EZ, et al. Using multispectral imaging for spoilage detection of pork meat. Food and Bioprocess Technology. 2013;**6**(9):2268-2279. DOI: 10.1007/s11947-012-0886-6

[10] Kamruzzaman M, Sun D-W, ElMasry G, Allen P. Fast detection and visualization of minced lamb meat adulteration using NIR hyperspectral imaging and multivariate image analysis. Talanta. 2013;**103**:130-136. DOI: 10.1016/j.talanta.2012.10.020

[11] Wu D, Shi H, Wang SJ, He Y, Bao YD, Liu KS. Rapid prediction of moisture content of dehydrated prawns using online hyperspectral imaging system. Analytica Chimica Acta. 2012;**726**:57-66. DOI: 10.1016/j.aca.2012.03.038

[12] Taghizadeh M, Gowen A, O'Donnell CP. Prediction of white button mushroom (*Agaricus bisporus*) moisture content using hyperspectral imaging. Sensing and Instrumentation for Food Quality and Safety. 2009;**3**(4):219-226. DOI: 10.1007/s11694-009-9088-y

[13] Rajkumar P, Wang N, Elmasry G, Raghavan GSV, Gariepy Y. Studies on banana fruit quality and maturity stages using hyperspectral imaging. Journal of Food Engineering. 2012;**108**(1):194-200. DOI: 10.1016/j.jfoodeng.2011.05.002

[14] ElMasry G, Wang N, ElSayed A, Ngadi M. Hyperspectral imaging for nondestructive determination of some quality attributes for strawberry. Journal

of Food Engineering. 2007;**81**(1):98-107. DOI: 10.1016/j.jfoodeng.2006.10.016

[15] Mendoza F, Lu RF, Ariana D, Cen HY, Bailey B. Integrated spectral and image analysis of hyperspectral scattering data for prediction of apple fruit firmness and soluble solids content. Postharvest Biology and Technology. 2011;**62**(2):149-160. DOI: 10.1016/j.postharvbio.2011.05.009

[16] Zhao J, Chen Q, Cai J, Ouyang Q. Automated tea quality classification by hyperspectral imaging. Applied Optics. 2009;**48**(19):3557-3564. DOI: 10.1364/AO.48.003557

[17] Kennard RW, Stone LA. Computer aided design of experiments. Technometrics. 1969;**11**(1):137-148. DOI: 10.1080/00401706.1969.10490666

[18] Yu K, Zhao Y, Li X, Shao Y, Zhu F, He Y. Identification of crack features in fresh jujube using Vis/NIR hyperspectral imaging combined with image processing. Computers and Electronics in Agriculture. 2014;**103**:1-10. DOI: 10.1016/j.compag.2014.01.016

[19] Polder G, Gerie WAM, Van DH. Calibration and characterization of imaging spectrographs. Journal of Near Infrared Spectroscopy. 2003;**11**(3): 193-210. DOI: 10.1255/jnirs.366

[20] Xing J, Bravo C, Jancsók P, Ramon H, De Baerdemaeker J. Detecting bruises on 'Golden delicious' apples using hyperspectral imaging with multiple wavebands. Biosystems Engineering. 2005;**90**(1):27-36. DOI: 10.1016/j.biosystemseng.2004.08.002

[21] ElMasry G, Wang N, Vigneault C. Detecting chilling injury in red delicious apple using hyperspectral imaging and neural networks. Postharvest Biology and Technology. 2009;**52**(1):1-8. DOI: 10.1016/j.postharvbio.2008.11.008

[22] Li H, Liang Y, Xu Q, Cao D. Key wavelengths screening using competitive adaptive reweighted sampling method for multivariate calibration. Analytica Chimica Acta. 2009;**648**:77-84. DOI: 10.1016/j.aca.2009.06.046

[23] Wu D, Nie PC, Cuello J, He Y, Wang ZP, Wu HX. Application of visible and near infrared spectroscopy for rapid and non-invasive quantification of common adulterants in spirulina powder. Journal of Food Engineering. 2011;**102**:278-286. DOI: 10.1016/j.jfoodeng.2010.09.002

[24] Wu D, Sun D-W. Potential of time series-hyperspectral imaging (TS-HSI) for non-invasive determination of microbial spoilage of salmon flesh. Talanta. 2013;**111**:39-46. DOI: 10.1016/j.talanta.2013.03.041

[25] Jespersen BM, Munck L. Cereals and cereal products. In: Sun DW, editor. Infrared Spectroscopy for Food Quality Analysis and Control. New York, USA: Academic Press; 2009. pp. 275-319. DOI: 10.1016/B978-0-12-374136-3.00011-0

Chapter 6

Use of Hyperspectral Remote Sensing to Estimate Water Quality

Mbongowo Mbuh

Abstract

Approximating and forecasting water variables like phosphorus, nitrogen, chlorophyll, dissolved organic matter, and turbidity are of supreme importance due to their strong influence on water resource quality. This chapter is aimed at showing the practicability of merging water quality observations from remote sensing with water quality modeling for efficient and effective monitoring of water quality. We examine the spatial dynamics of water quality with hyperspectral remote sensing and present approaches that can be used to estimate water quality using hyperspectral images. The methods presented here have been embraced because the blue-green and green algae peak wavelengths reflectance are close together and make their distinction more challenging. It has also been established that hyperspectral imagers permit an improved recognition of chlorophyll and hereafter algae, due to acquired narrow spectral bands between 450 nm and 600 nm. We start by describing the practical application of hyperspectral remote sensing data in water quality modeling. The surface inherent optical properties of absorption and backscattering of chlorophyll a, colored dissolved organic matter (CDOM), and turbidity are estimated, and a detailed approach on analyzing ARCHER data for water quality estimation is presented.

Keywords: water quality, field spectroscopy, ARCHER, chlorophyll a, colored dissolved organic matter, turbidity, total phosphorus, nitrogen

1. Introduction

Water is one of the valuable and essential resources of life on earth. There is ever-increasing stress on water resources, and as population increases, there is an ever-increasing pressure placed on water resources [1–3]. Several nations depend on water resources for economic growth [4]. Water serves as a source of food, income, and livelihood for many [4, 5]. Equally, important information on resources that support life in an ecosystem is delivered by the quality of surface water [6]. An increase in water pollution deteriorates water quality and also threatens human health, aquatic ecosystem balance, economic development, and social prosperity [7, 8].

Supportable water resources management requires continuous and accurate monitoring. Satellite observations [1, 9] have provided data for such tracking for several years [10] and have served at a time- and the cost-effective way to carry out large-scale monitoring [11, 12]. Water pollution is an important environmental issue, further limiting the availability of water for human and environmental use [1, 13]. Though nutrients are indispensable for plant and animal growth and nourishment, an excess of some nutrients in water can disturb the river [1, 14, 15]. Excellent and clear water is imperative to the plants and animals that live in any watershed.

A significant difficulty in assessing surface water quality is identifying the sources of pollutants and the contribution of the parameters/variables that explain water quality variation [1, 6, 16–18]. Determining the conditions and parameters of water quality is one of the significant advantages of hyperspectral remote sensing technologies. Hyperspectral reflectance technology has been broadly used to examine and monitor the water quality conditions of many open water aquatic ecosystems [19, 20]. Hyperspectral remote sensing has been used to characterize algal blooms [21] and assess ammonia dynamics for wetland treatments [22, 23]. Tilley et al. [23] also developed remotely sensed hyperspectral signatures of macrophytes to monitor changes in wetland water quality predictors of total ammonia concentrations [24] Hyperspectral remote sensing has similarly been used to determine water quality parameters like temperature, chlorophyll a, total suspended solids [25, 26], total phosphorus [27, 28], and turbidity; Lillesand et al. [29] and Lathrop and Lillesand [30] studied lakes and reservoirs, estuaries [31, 32], and tropical coastal areas [33, 34]. Other water quality studies on monitoring surface water bodies in different parts of the world (e.g., [35–40]) have all been interested in modeling and development of concentration distribution maps for different water quality parameters based on its reflectance characteristics. Algal concentrations in water through hyperspectral remote sensing images have been undertaken in the estimation of chlorophyll that is then used as an estimate for monitoring algal content and hence water quality. This approach has been adopted because wavelengths corresponding with the peak reflectance of blue-green and green algae are close together; it is harder to differentiate between them [19, 41, 42]. Hakvoorth et al. [43], however, demonstrate that hyperspectral imagers permit for improved detection of chlorophyll and hereafter algae, as a result of acquired narrow spectral bands between 450 nm and 600 nm [20, 44].

1.1 Remote sensing for water quality

The spectral signature changes in the water can be measured and relate them to empirical or analytical models to a water quality parameter through remote sensing techniques [25, 45]. Since the 1960s, the earth's resources have been monitored from space by the National Aeronautics and Space Administration (NASA) with multispectral scanners, which collect data sets in about 5–10 bands of relatively large bandwidths (70–400 nm) [10, 46]. The spectral resolution of data from the multispectral scanners was limited, inadequately evaluating water quality and starting in the mid-1980s. Hyperspectral remote sensing with a higher spectral resolution (i.e., 224 bands) and 30 meters in spatial resolution covering wavelengths from the 400–2500 nm "in the visible and near-infrared bands of the spectrum" (Field Assessment of a Fiber Optic Spectral Reflectance System http://horttech.ashspublications.org/content/6/1/73.full.pdf) became available for earth sciences including water quality monitoring. Some of these hyperspectral sensors include FTHSI on MightySat II, Hyperion on NASA EO-1, airborne visible/infrared imaging spectrometer (AVIRIS), Airborne real-time cueing hyperspectral enhanced reconnaissance (ARCHER), Hyperspectral Digital Imagery Collection Experiment (HYDICE), PROBE-1, Compact Airborne Spectrographic Imager (CASI), and HyMap. The ARCHER sensor, which is of interest to this research, is used to estimate the water quality parameters. The very high spectral resolution of hyperspectral sensors gives them the advantage over multispectral sensors in facilitating exceptional differentiation of objects based on their spectral response in the narrow bands [10, 47]. This spectral information has made hyperspectral sensor data very useful in estimating dissolved organic matter, chlorophyll, and total suspended matter concentrations from optical remote sensing technologies [43, 48, 49]. Our objective is to review the literature of water quality as it relates to remote sensing, water quality modeling, and data fusion.

The application of hyperspectral remote sensing techniques to water resource problems is proving to be the most in-depth way of examining spatial, spectral, and temporal variations to derive more accurate estimates of information required for water resource applications [19]. This emergence offers the capability of covering large areas on a real-time scale to directly monitor and characterize environmental pollutants entering a body of water. Addressing the problem of colored dissolved organic matter (CDOM), Nelson and Guarda [50], in the South Atlantic Bight, and Vodacek et al. [51], in the Mid-Atlantic, examined the visible absorption spectra and characteristics of particulate and dissolved materials. Both studies demonstrated that colored dissolved organic matter comes mostly from riverine runoff, and it is also widespread and abundant in natural waters, which have a significant portion of the dissolved organic matter (10–90%), and influences water-leaving radiances [52]. Another chlorophyll retrieval study by Fell et al. [53] used chlorophyll algorithms to describe coastal properties in the Monterey Bay through hyperspectral remote sensing. Using a composite AVIRIS to examine marine environment changes, the Sea-viewing Wide Field-of-view Sensor (SeaWiFS) algorithm was applied to derive chlorophyll information. The study showed the importance of high spatial resolution in representing the coastal ocean in the Monterey Bay, though additional research using higher hyperspectral resolution on the phytoplankton pigment spectral absorbance was recommended.

Kirkpatrick et al. [54] indicate that a considerable portion of the organic carbon in the oceans is found as dissolved organic matter (DOM) and a better understanding of the distribution and dynamics of DOM is necessary for understanding global carbon cycles. The authors also demonstrate that CDOM is often present in concentrations sufficient to affect the color of lakes, estuaries, and nearshore coastal waters, although other studies have shown that CDOM absorption does not correlate with chlorophyll a [55, 56]. Brando and Dekker [57] used spectroscopy to test for its capabilities over a range of water targets in eastern Australia using open ocean flushing and a combination of turbid and humic river inputs, to determine the water quality of the bay. Integrated atmospheric and hydro-optical radiative transfer models (MODTRAN- 4, Hydrolight) were developed to estimate the underwater light field. A matrix inversion approach was used to retrieve chlorophyll a, dissolved organic matter, and suspended matter concentrations. The research demonstrated that Hyperion has enough sensitivity to map optical water quality concentrations of total suspended matter, dissolved organic matter, chlorophyll, and concurrently the complex waters of estuarine and coastal systems of the Moreton Bay. The results obtained from this retrieval were comparable to those estimated in the field campaigns, which were coincident with Hyperion overpasses. [38], in a similar study, collected three sets of remote sensing and ground-truth data to evaluate the correlations between reflection data and water quality analyses to develop optical indicators of water quality constituents. Imagery and field reflectance data and water quality samples were collected in the river in 1999 concurrently, and 2001, in southwest Ohio, and results showed a correlation between the spectral data and water quality parameters.

Brezonik et al. [36] used Landsat-based remote sensing to characterize chlorophyll a, total suspended sediments (TSS), turbidity, and Secchi disk transparency (SDT) of lake water quality. All three variables demonstrated a high correlation with each other, and all act as direct or indirect measures of algal abundance in

Minnesota lakes. This study also showed that chl-a and turbidity could be estimated from Landsat data if the near-contemporaneous ground measurements are available for calibration. Also, Kneubühler et al. [58], in evaluating total chlorophyll content (TCHL) concentration, used spectral reflectance data measured at 1 m above the water surface with a handheld field spectroradiometer and applied the semi-analytical algorithms. The results proved to be valuable for an enormous range of observed TCHL concentrations (0–460 μg/L), high r^2, and low mean deviations. Dingtian et al. [59] used hyperspectral remote sensing images and field reflectance measurements with Field spec, to characterize chl-a and suspended solids in Taihu Lake, China. Their results showed the relationship between chl-a and wavelengths in Taihu Lake in different seasons, with an average correlation coefficient of more than 0.65. This research showed success in the application of hyperspectral remote sensing in retrieving chl-a and suspended solid concentrations.

Giardino et al. [60] used hyperspectral data to map chlorophyll a and tripton concentrations in Lake Garda based on the forward and inverse bio-optical modeling. The research demonstrated that Hyperion-derived levels were on average comparable to in situ data for chlorophyll a. The authors, however, mentioned that the same analysis was more complicated for tripton since some incompatibilities of methods occurred. This study demonstrated that the spatial and spectral resolutions of Hyperion and the capability of physics-based approaches were considered highly suitable, although more research was necessary to address the compatibilities of methods for monitoring waterbody features with a high rate of wind or wave-driven change. This study also showed that procedures used can be transferred to other water bodies if the optical characterization of the water body is known and information about atmospheric properties during the satellite overpass is accessible.

Equally, Giardino et al. [61] used satellite data and field spectrometer data to estimate chl-a as an indicator of the trophic level and CDOM in the Curonian Lagoon. A PANalytical handheld spectroradiometer in situ Rrs spectra can be used to parameterize a semi-empirical algorithm in retrieving chl-a concentrations and validate the performances of two atmospheric correction algorithms, to build a bond ratio algorithm for chl-a and to validate MERIS-derived maps. Results from this combined in situ and calibration study confirmed the hypertrophic/dystrophic conditions of the Curonian Lagoon.

Santini et al. [62], to analyze colored dissolved organic matter, used hyperspectral remote sensing techniques ranging from empirical algorithm to complex physics-based models to retrieve water quality constituent. With the empirical approach, acceptable results for the CDOM concentrations were returned. The study also showed a correlation index of over 0.82, between the laboratory CDOM concentrations and model output. The study showed that the physical model could be used to retrieve simultaneously of chlorophyll and the total suspended matter concentrations. Another research studying the relationship between suspended sediments and reflectance has been demonstrated to rely on physical and optical characteristics of sediment type and sensor zenith angle [63], and the properties of scattering and absorption of sediment type affect water reflectance [64].

Xiao et al. [65] explored the potential of in situ hyperspectral remote sensing for estimating chlorophyll a and phycocyanin concentrations of a water body. In situ measurements of the lake surface reflectance at the five sites were examined using PANalytical FieldSpec3 spectroradiometer to investigate the relationship between PANalytical-based reflectance data and chlorophyll a and phycocyanin concentrations at different depths of water. The study shows significant correlations between lake surface reflectance and chlorophyll a and phycocyanin concentrations in upper mixed surface waters (0 to 1 m depth) at these five sites. Hommersom et al. [66] also used PANalytical field spec to carry out measurements in the central basin of Lake Vänern, and matrix inversion algorithms were used to derive parameters such as the concentrations of chl-a and suspended particulate matter (SPM) and the absorption by colored dissolved organic matter at 440 nm. Maltese et al. [67] retrieved turbidity from MODIS data, and PANalytical handheld spectrometer was used to obtain underwater irradiances at 11 depths from just below the water's surface, up to 5.5 meters. In situ data, acquired during the spring and summer, were used to enhance the retrieval of water surface nephelometric turbidity locally through satellite images.

2. Remote sensing of water quality analysis approaches

2.1 Empirical approach and analytical methods

There exist two main approaches to examining water quality from remotely sensed data: the (semi-) empirical approach and the (semi-) analytical method [60, 68–72]. The most common are the semi-empirical and empirical approaches where water quality is determined by statistical relationships between measured spectral properties (reflectance) and the measured water quality parameter of interest [72]. Ocean color derivation algorithms for chlorophyll a concentration have applied this approach to high correlations between chl-a and the blue and green spectral regions (chl-a has absorption maxima at 430–450 nm and 660–680 nm (nanometers)) (Reif [73]). However, Dall'Olmo and Gitelson [74] have illustrated that these spectral regions typically do not work, and this problem has been fixed by subtracting the contributions of other factors on reflectance nearby the peak at 670 nm with a three-band reflectance model [75, 76].

With the use of empirical approaches, statistical regressions are recognized among reflectance values extracted from the image with synchronized in situ water quality measurements for correlation and validation well for retrieval of chl-a in waters with increased turbidity and overlapping absorption of dissolved organic matter and tripton [73, 76]. Using this method wavelengths are naturally evaluated and selected from regions in the spectrum in which absorption and reflectance are strongly impacted by the parameter of interest [68]. Band ratio algorithms between a reflectance peak near 700 nm and an absorption peak (red chl-a absorption band) around 670–680 nm have been developed for turbid water environments to retrieve chlorophyll [73]. Though the empirical approach has shown some success, it has the disadvantages that they require in situ samplings for testing and validation and they tend to be scene dependent, to apply locally to the explicit data from which they were derived [60, 68, 72, 77].

To solve this problem, analytically and semi-analytical approaches that mention modeling that is more complex where water parameter concentrations are related physically to the measured reflectance spectra by evaluating their absorption and scattering coefficients at multiple wavelengths are necessary to take care of the problems [73]. This method establishes sophisticated radiative transfer equations, relationships between water reflectance and the concentration of constituents and their specified inherent optical properties (SIOPS) [60, 68, 70, 72]. Using the analytical approach, the radiative transfer equation is inverted to determine water quality parameters, and several inversion procedures have been established for this purpose [78, 79] and have been revealed to optimize unknown parameters when measured input does not exist [60, 62, 78].

The inversion approach has been vital to separate bottom reflectance from water column spectra, in superficial waters where the water-leaving radiance/reflectance possibly encompasses some spectral evidence from the bottom reflectance and in the water column [73, 80]. Using the simple methods like the empirical method, optically, shallow water can result in an overestimation of water column constituents caused by high reflectance values primarily from the bottom reflectance [81]. Comparing empirical and analytical approaches, it can be noted that analytical and semi-analytical methods are preferred for subsequent reasons: (1) they can be used to estimate both optically profound and shallow water optical properties, and the bottoms of optically shallow waters with physics-based modeling; (2) the approach does not require in situ water quality measurements to model, resulting in its independence; and (3) analytical and semi-analytical methods can be applied regionally in multiple lakes, reservoirs, and rivers with varied circumstances. Notwithstanding these benefits, nonetheless, they are computationally intensive and more expensive and difficult to use, thus requiring information of the inherent optical properties of the water body [73]. This research relies on the analytical approach to analyzing spectroscopic data.

2.2 Use of hyperspectral remote sensing methods and standard water quality approach in measuring the water quality parameters

Although the standard methods provide accurate measurement for a point in time and space, spatial or temporal view of water quality required for precise assessment of large water systems is usually not available [72]. It is necessary to integrate the use of calibrated image data with field spectral measurements to solve this problem, so as entirely to deploy the spatial and spectral information of hyperspectral remote sensing data. Hyperspectral images are critical for the water quality assessments where field data collection is planned to coincide with flight overpasses followed by the retrieval of the apparent and inherent optical properties of the basin or watershed of interest.

An in situ sampling water quality survey for nutrients is necessary at multiple sites in the study area, using the EPA-approved quality control/quality assurance procedures. A sample collected procedure is required, and we recommend 15 to 20 samples from each sampling area separated by at least 100 m from each other; using handheld spectrometer and paying particular attention to just the deep portions of the river for sample collection, above surface water reflectance was also measured. In situ data for chlorophyll a and other nutrients of interest can also be obtained from water quality databases, which contain data for fixed monitoring stations throughout the watershed of interest.

Using the handheld spectrometer to measure all the relevant quantities from above the surface, three types of measurements were carried out at each sampling site with the spectrometer: total upwelling radiance (LT), downwelling sky radiance (LSky), and “gray-card” radiance (LG, 3) reflected from a diffuse reflector (Spectralon®) [71]. All measurements were carried out at about 2:30 pm (local time), under clear skies, minor cloud cover, a wind speed of 4 m s^{-1}, and very calm water, at roughly 0.5 m above the water surface using a canoe. The above-water reflectance needs to be measured at 40° from the nadir and 90° from the azimuth and the sky reflectance measured in the same plane as the water, except for the angle from the zenith, which was 40°. To determine the downwelling irradiance, the Spectralon is assumed to have a Lambertian reflector in which, Ed = πLG/R, where LG is the average of the four grayscale scans and RG the reflectance of the diffuse reflector (~10%) [71].

2.3 Hyperspectral image processing

For the quantitative assessments of water quality parameters, detectable from hyperspectral data, data preprocessing is required by performing robust corrections for atmospheric effects of adjacency effects and those effects occurring at the water surface level (sunglint, specular reflection of direct irradiance, and diffuse skylight).

Hyperspectral imagery requires an atmospheric correction to retrieve the surface reflectance from remotely sensed imagery by removing the atmospheric effects such as water vapor and other trace gasses. In an atmospheric correction, the radiance values are transformed into reflectance data to obtain water reflectance by removing surface reflectance [82], measuring the fraction of radiation reflected from the surface [83]. This procedure is particularly important for quantitative image analysis or change detection using hyperspectral data; image calibration is essential for remote sensing (**Figure 1**) to convert the instrument's digital numbers (DNs) to a substantial value to correct atmospheric instrument effect.

Image-driven empirical correction procedures have been suggested [57, 84, 85] for use with the Hyperspectral Imager for the Coastal Ocean (HICO), airborne visible/infrared imaging spectrometer (AVIRIS), Compact Airborne Spectrographic Imager (CASI-2), and Hyperion [86]. The empirical correction approach is based on the facts that clear ocean waters have water-leaving reflectance above 800 nm close to zero and sunglint and cirrus reflectance in the 400–1000 nm region. In this dissertation, we use the empirical line approach, which is an atmospheric correction method that serves as an alternative to radiative transfer modeling approaches [87]. This method calculates the empirical relation between radiance and reflectance using a dark and a bright target, well-characterized by field and image spectra. Our targets were measured in the area during data collection for optimal representation.

This method has been applied to correct both land and ocean data [88] and has shown great success with both coarser spatial resolution satellite sensor data and airborne data approaches [87]. This technique is only suitable for regional data correction where reflectance properties of bright and dark targets such as sand and water over uniform areas are measured coincidentally with the aircraft or airborne overpass [89].

A minimum of two known materials is required to use this method to carry out the calibration, and selecting one bright object and one dark object is also crucial for this exercise. This calibration method is recommended to use on two targets; however, using more targets will better estimate the relationship between target reflectance and at-sensor radiance [87, 88]. Using the image and field spectra, the two targets are regressed linearly against the reflectance spectra measured on the

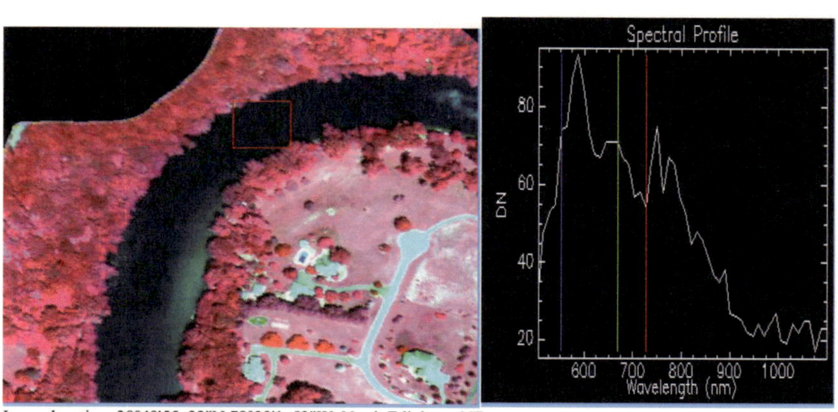

Image location: 38°49'55. 32"N 78°33'1. 53"W, North Edinburg, NF

Figure 1.
ARCHER color composite (RGB of 726 nm, 668 nm, and 551 nm) for areas around Edinburg, the NorthFork of Shenandoah River in Virginia, USA. Image location: 38°49'55. 32"N 78°33'1. 53"W, North Edinburg, NF.

field to derive the gain and offset coefficients [88]. Once the gains and offsets are obtained, they are then applied to the entire image to derive surface reflectance, by producing reflectance values that are comparable to field measured values [88] (**Figures 1** and **2**). The empirical line method uses the following equation to calculate the gains and offsets:

Reflectance (field spectrum) = gain* radiance (input data) + offset.

Remote sensing data is also impeded by the effect of wave-induced sun glint [90], and this has become a limiting factor in estimating water quality efficiently from airborne data with high accuracy. The environmental and atmospheric effects resulting in inaccuracies in remote sensing classification results remain a growing concern in remote sensing classification [91]. For an adequate estimation of water quality with remote sensing data that is void of inaccuracies, the sun glint needs to be examined. After performing atmospheric correction on our image, sunglint removal was required to correct atmospheric effects on the visible wavelength region (0.45–0.69 μm). The sunglint is the specular reflection of sunlight directly transmitted from the air-water interfaces [92]. Under clear skies and irregular water surface, specular reflectance can result in sun glint on the image, which reduces the accuracy of retrievals [93]. The sunglint often occurs on an image when the orientation of the water surface is directly reflected toward the sensor as a function of the position of the sun, the viewing angle, and the state of the water surface [92].

These circumstances have resulted to the more excellent specular reflection of light from water "than the water-leaving radiance from the sub-surface features." The necessity to remove the sun glint contribution for better image classification or information retrieval has been recognized by several researchers [90]. The approach adopted for this research estimates the amount of glint in the image by using data from the near-infrared (NIR), with the assumption that water-leaving radiance is negligible in this part of the spectrum, and any NIR signal left after atmospheric correction is undoubtedly from the sunglint. A relationship is established between the NIR and glint radiance while using the spectrum of the deep-water part of the image [92]. We use the shallow water sunglint removal approach that assumes that all the radiance from the NIR reaching the sensor is from atmospheric scattering and surface reflection, and any signal at the NIR after atmospheric correction is sunglint [92].

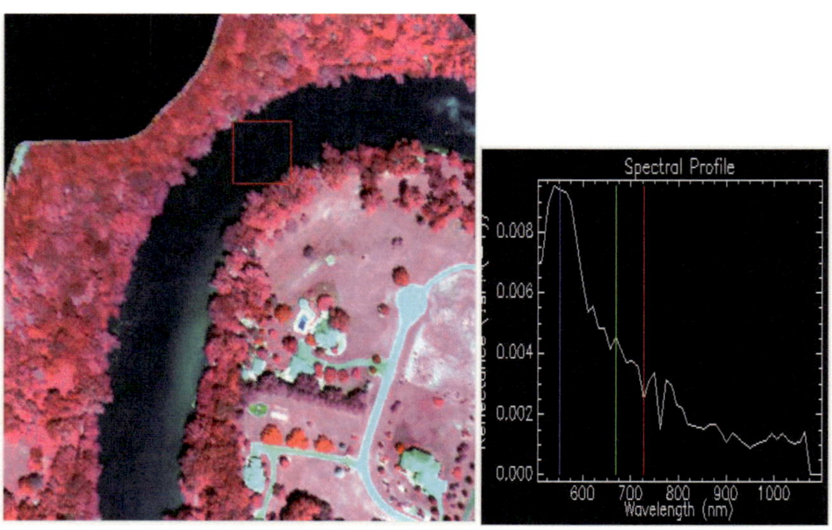

Figure 2.
Atmospherically corrected ARCHER using empirical line calibration approach with a color composite of RGB 726 nm, 668 nm, and 551 nm for areas around Edinburg (above), for North Fork of the Shenandoah River taken on July 12, 2014. Image location: 38°49'55. 32"N 78°33'1. 53"W, North Edinburg, NF.

3. Sunglint background and removal approach

There are five critical processes through which a remote sensing detector receives radiance reaching it, as shown in **Figure 3** from Kay et al. [92].

Several approaches have been proposed for glint correction for estimating the contribution of glint to the " the sensor reaching radiance, and then subtract it from the received signal" [92]. Hochberg et al. [94] proposed a sunglint removal method, which assumed that the NIR brightness is only made up of sunglint and a spatially constant ambient NIR component. This method also believes that the sunglint present in the visible band is linearly related to the brightness of the NIR band. However, all two assumptions were proven weak because the first assumption models a constant ambient NIR brightness, which is removed from all pixels during analysis, and secondly, only two pixels are used to establish a linear relationship assumption. Selecting only one bright and one dark pixel could result to a bright pixel chosen from the land, which necessitated masking for results from this method to be efficient, and, this makes it very difficult and time-consuming. Thus, the difficulty of being able to identify an appropriate bright pixel can result in significant errors, which undermine the effectiveness of the method proposed by Hochberg et al. [94].

Hedley et al. [95], after acknowledging how sensitive this approach was to outlier pixel, proposed a revised method in which glint intensity is obtained using several pixels rather than two to establish a linear relationship between regression between the NIR and visible bands to allow sunglint contribution removal [90]. Hedley et al. [93] proposes using single or several regions on the image where sunglint is evident with consistent spectral brightness. The linear regression uses NIR brightness (x-axis) against the visible band's intensity (**Figure 4**) of all the selected pixels.

As recommended by Hedley et al. [93], the first step is to select the minimum NIR brightness NIR Min deep-water pixels having a variety of glint intensities from which a sample is calculated. The next step in deglinting the image is to use each visible spectrum (VIS) Band i and perform a linear regression on the NIR pixel

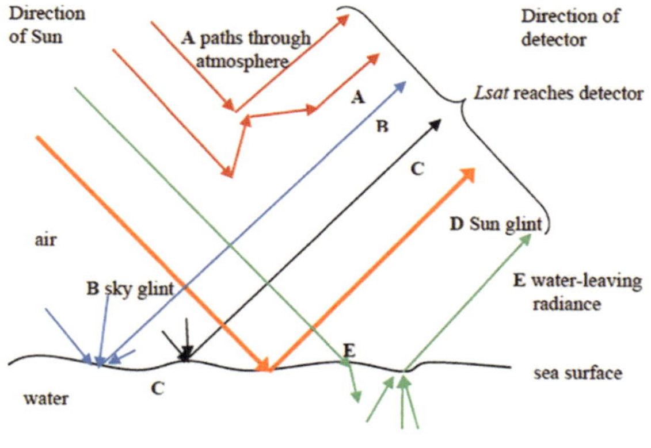

Figure 3.
Diagram showing routes by which light can arrive at a remote sensing detector from Kay et al. (2005). (A) Molecules or aerosols scattering in the atmosphere, which is either single or multiple. (B) Surface-water scattering from the atmosphere followed by reflection to the detector—known as "sky glint." (C) Whitecaps reflections from the sea surface. (D) Surface-water specular reflection directly transmitted from the sun to the atmosphere to the surface and from the surface to the detector—also called "sun glint." (E) Atmosphere and air-water interface transmission, which is followed by reflection or scattering below the water surface and transfer back to the detector through the atmosphere.

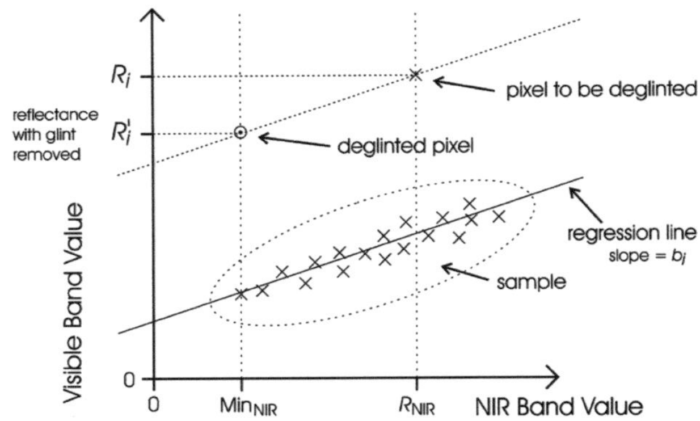

Figure 4.
Graphical interpretation of sunglint correction method from Hedley et al. [93].

brightness R NIR against the pixel value of VIS band R i. A user-based selection process is used to collect the samples, and land or cloud masking is not necessary. The product of slope b_i and R_{NIR} minus Min_{NIR} is subtracted from R_i to obtain the pixel R_i with glint removed using the following equation:

$$R'_i = R_i - b_i(R_{NIR} - Min_{NIR}) \tag{1}$$

where b_i is the regression slope.

R_i is the visible band.

R_{NIR} is the NIR pixel value.

Min_{NIR} is the ambient NIR value, which is NIR pixel with no sunglint, which is either estimated from the figure above or from the entire image, and it is less prone to outliers caused by nonoptically deep pixels.

The result of the sunglint corrects brightness in band i, by minimizing outlier effects caused by surface objects [92]. This approach can be applied on either before or after atmospheric correction since it works entirely on the relative magnitude of values, and the pixel units are not very necessary for image deglinting. We initially corrected out the image with the empirical line method before removing the sunglint. It should, however, be mentioned that, if there are variations in the atmosphere properties, this will also affect the regression slope, thus making glint effect to be confounded [92]. As outlined by Hedley et al. [93], this approach is attained in four steps:

Step-by-step implementation

1. Image is radiometrically corrected.

2. Area of the image displaying a range of sun glint, with a more or less homogeneous surface, is selected. The minimum NIR brightness value is determined.

3. The newly created region of interest is used as a subset to create a new image with only the glare pixel subset and all image bands saved individually in ASCII. A linear regression of NIR brightness (x-axis) against the visible band (y-axis) is performed using the selected pixels in Excel to remove the sunglint from each band. The output of interest from the linear regression analysis for each band is the slope, which is called bi in the equation above.

4. To individually glint each band i or all pixels in the image, the product of bi and NIR brightness of the pixel (minus Min_{MIN}) subtracted the pixel value in band i as illustrated in Eq. (1).

4. Conclusion

The combination of several hundred spectral bands in a single acquisition has been made feasible by hyperspectral systems, which produce more detailed spectral data. Before advances in hyperspectral remote sensing, the multispectral imagery was the only data source in land and water observational remote sensing from airborne and spacecraft operations since the 1960s [10]. However, multispectral remote sensing data were only collected in three to six spectral bands in a single observation from the visible near-infrared and shortwave infrared regions of the electromagnetic spectrum, making it challenging to examine water quality from this data source. The present chapter covered hyperspectral remote sensing data analysis using field spectrometer data and remote sensing of water quality. Research has shown that remote sensing, GIS, and hydrological models can be integrated to solve hydrological problems [96, 97]. Here we review relevant literature on research in hyperspectral remote sensing that examines water quality parameters like suspended sediments, turbidity, chlorophyll a, and total phosphorus as investigated by numerous researchers. Unique characteristics of hyperspectral remote sensing data are introduced. This chapter shows that field observations/ spectroscopy, and water quality modeling is very instrumental in the accuracy of remote sensing analysis. We also presented the methodology for the study of visible to infrared hyperspectral remote sensing data from ARCHER aircraft and data collected with a handheld field portable spectroradiometer, to retrieve and establish a relationship between water quality parameters like chlorophyll a, colored dissolved organic matter, turbidity, phosphorus, and nitrogen in the Shenandoah River Basin.

Author details

Mbongowo Mbuh
Department of Geography and Geographic Information Science, University of North Dakota, USA

*Address all correspondence to: mbongowo.mbuh@und.edu

References

[1] Mbuh MJ, Houser PR, Heidari A. Water quality estimation using combined water chemistry and field spectroscopy in the Shenandoah River, Virginia. International Journal of Applied Geospatial Research (IJAGR). 2016;7(2):14-37. DOI: 10.4018/ijagr.2016040102

[2] Mcgwire K, Minor T, Fenstermaker L. Hyperspectral mixture modeling for quantifying sparse vegetation cover in arid environments. Remote Sensing of Environment. 2000;**72**:360-374

[3] Hinrichsen D, Tacio H. The Coming Freshwater Crisis is Already Here. Population and Water. 2019. p. 26. Available from: https://www.wilsoncenter.org/sites/default/files/popwawa2.pdf

[4] UNESCO. Water for a sustainable world. Paris: UNESCO.Unesco 2015a, Water for a sustainable world. pdf. 2015. Available from: https://sustainabledevelopment.un.org/content/documents/1711Water%20for%20a%20Sustainable%20World.pdf

[5] Cook SE, Fisher MJ, Andersson MS, Rubiano J, Giordano M. Water, food and livelihoods in river basins. Water International. 2009;**34**(1):13-29. DOI: 10.1080/02508060802673860

[6] Mustapha A, Aris AZ. Spatial aspects of surface water quality in the Jakara Basin, Nigeria, using chemometric analysis. Journal of Environmental Science and Health, Part A. 2012;**47**(10):1455-1465

[7] Milovanovic M. Water quality assessment and determination of pollution sources along the Axios/Vardar River, Southeast Europe. Desalination. 2007;**213**:159-173

[8] Verma S. Analysis of Water Quality by Physico-Chemical Parameters in Fateh Sagar Talab in Bagar, Dist. Of Jhunjhunu (Raj.), India; 2015

[9] Sheffield J, Wood EF, Pan M, Beck H, Coccia G, Serrat-Capdevila A, et al. Satellite remote sensing for water resources management: Potential for supporting sustainable development in data-poor regions. Water Resources Research. 2018;**54**(12):9724-9758

[10] Landgrebe D. Some fundamentals and methods for hyperspectral image data analysis. In: Systems and Technologies for Clinical Diagnostics and Drug Discovery II. Vol. 3603. Society of Photo-Optical Instrumentation Engineers; 1999. 6 pp

[11] Okin SG, Roberts AD, Murray B, William J, Okin JW. Practical limits on hyperspectral vegetation discrimination in arid and semiarid environments. Remote Sensing of Environment. 2001;**77**:212-225

[12] Wang X, Yang W. Water quality monitoring and evaluation using remote-sensing techniques in China: A systematic review. Ecosystem Health and Sustainability. 2019;**5**(1):47-56

[13] Cosgrove WJ, Loucks DP. Water management: Current and future challenges and research directions. Water Resources Research. 2015;**51**(6):4823-4839

[14] Bhateria R, Jain D. Water quality assessment of lake water: A review. Sustainable Water Resources Management. 2016;**2**(2):161-173

[15] David MK, Dennis HR. Nutrients in the Nation's Waters—Too Much of a Good Thing? (Report No. 1136). Vol. 1136. U.S. Geological Survey Circular; 1996. 24p. DOI: 10.3133/cir1136

[16] Dunca A-M. Water Pollution and Water Quality Assessment of Major Transboundary Rivers from Banat (Romania) [Research Article]. 2018

[17] Fataei E. Assessment of surface water quality using principle component analysis and factor analysis. World Journal of Fish and Marine Sciences. 2011;**3**:159-166

[18] Meng W, Zhang N, Zhang Y, Zhang BH. Integrated assessment of river health based on water quality, aquatic life, and physical habitat. Journal of Environmental Sciences. 2009;**21**:1017-1027

[19] Govender M, Chetty K, Bulcock H. A review of hyperspectral remote sensing and its application in vegetation and water resource studies. Water SA. 2006;**2007**:33(2)

[20] Mbuh MJ. Optimization of airborne real-time cueing Hyperspectral enhanced reconnaissance (ARCHER) imagery, in situ data with chemometrics to evaluate nutrients in the Shenandoah River, Virginia. Geocarto International. 2018;**33**(12):1326-1349. DOI: 10.1080/10106049.2017.1343395

[21] Stumpf RP. Applications of satellite ocean color sensors for monitoring and predicting harmful algal blooms. Human and Ecological Risk Assessment. 2001;**7**:1363-1368

[22] Kang W, Chai H, Xiang Y, Chen W, Shao Z, He Q. Assessment of low concentration wastewater treatment operations with dewatered alum sludge-based sequencing batch constructed wetland system. Scientific Reports. 2017;**7**(1):17497

[23] Tilley DR, Ahmed M, Son JH, Badrinarayanan H. Hyperspectral reflectance of emergent macrophytes as an indicator of water column ammonia in an oligohaline, subtropical marsh. Ecological Engineering. 2003;**21**:153-163

[24] Guo M, Li J, Sheng C, Xu J, Wu L. A review of wetland remote sensing. Sensors. 2017;**17**(4):777. DOI: 10.3390/s17040777

[25] Gholizadeh MH, Melesse AM, Reddi L. A comprehensive review on water quality parameters estimation using remote sensing techniques. Sensors (Basel, Switzerland). 2016;**16**(8). DOI: 10.3390/s16081298

[26] Vos RJ, Hakvoort JHM, Jordans RWJ, Ibelings BW. Multiplatform optical monitoring of eutrophication in temporally and spatially variable lakes. Science of the Total Environment. 2003;**312**:221-243

[27] Gao Y, Gao J, Yin H, Liu C, Xia T, Wang J, et al. Remote sensing estimation of the total phosphorus concentration in a large lake using band combinations and regional multivariate statistical modeling techniques. Journal of Environmental Management. 2015;**151**:33-43

[28] Koponen S, Pulliainen J, Kallio K, Kallikainen M. Lake water quality classification with airborne hyperspectral spectrometer and simulated MERIS data. Remote Sensing of Environment. 2002;**79**:51-59

[29] Lillesand TM, Johnson WL, Deuell RL, Lindstrom OM, Meisner DE. Use of Landsat data to predict the trophic state of Minnesota Lakes. Photogrammetric Engineering and Remote Sensing. 1983;**49**(2):219-229

[30] Lathrop RG, Lillesand TM. Monitoring water quality and river plume transport in Green Bay, Lake Michigan with SPOT-1 imagery. Photogrammetric Engineering and Remote Sensing. 1989;**55**(3):349-354

[31] Harding LW, Itsweire EC, Esaias WE. Algorithm development for recovering chlorophyll concentrations int Chesapeake Bay using aircraft remote sensing 1989-91. Photogrammetric Engineering and Remote Sensing. 1995;**61**:177-185

[32] Zhang Y, Giardino C, Li L. Water optics and water colour remote sensing. Remote Sensing. 2017;**9**(8):818

[33] Devlin MJ, Petus C, Da Silva E, Tracey D, Wolff NH, Waterhouse J, et al. Water quality and river plume monitoring in the great barrier reef: An overview of methods based on ocean colour satellite data. Remote Sensing. 2015;7(10):12909-12941

[34] Ruiz-Azuara P. Multitemporal analysis of "simultaneous" Landsat imagery (MSS and TM) for monitoring primary production in a small tropical coastal lagoon. Photogrammetric Engineering & Remote Sensing. 1995;**61**(2):877-198

[35] Bagheri S, Dios RA. Utility of hyperspectral data for bathymetric mapping in a turbid estuary. International Journal of Remote Sensing. 1998;**19**(6):1179-1188

[36] Brezonik P, Menken KD, Bauer M. Landsat-based remote sensing of lake water quality characteristics, including chlorophyll and colored dissolvedorganic matter (CDOM). Lake and Reservoir Management. 2005;**21**(4):373-382

[37] Goodenough A. Evaluating water quality monitoring with hyperspectral imagery [Senior Research Thesis]. Rochester Institute of Technology: Center for Imaging Science; 2001

[38] Shafique NA, Fulk F, Autrey BC, Flotemersch J. Hyperspectral remote sensing of water quality parameters for large rivers in the Ohio River Basin. In: Proceedings of the 1st Interagency Conference on Research in the Watersheds. Benson, AZ, USA; 27-30 October 2003

[39] Usali N, Ismail MH. Use of remote sensing and GIS in monitoring water quality. Journal of Sustainable Development. 2010;**3**(3):228. DOI: 10.5539/jsd.v3n3p228

[40] Yang S, Ding P, Zhu J, Zhao Q, Mao Z. Tidal flat morphodynamic processes of the Yangtze estuary and their engineering implications. China Ocean Engineering. 2000;**14**:307-320

[41] Gokul EA, Shanmugam P. An optical system for detecting and describing major algal blooms in coastal and oceanic waters around India: Detecting algal blooms. Journal of Geophysical Research, Oceans. 2016;**121**(6):4097-4127

[42] Mcilwaine B, Casado MR, Leinster P. Using 1st derivative reflectance signatures within a remote sensing framework to identify macroalgae in marine environments. Remote Sensing. 2019;**11**(6):704

[43] Hakvoorth H, deHaah J, Jordan R, Vos R, Rijkeboer MP. Towards airborne remote sensing of water quality in the Netherlands-validation and error analysis. Journal of Photogrammetry and Remote Sensing. 2002;**57**:171-183

[44] Gao B-C, Li R-R. FVI—A floating vegetation index formed with three near-ir channels in the 1.0-1.24 μm spectral range for the detection of vegetation floating over water surfaces. Remote Sensing. 2018;**10**(9):1421

[45] Shafique NA, Autrey BC, Fulk F, Cormier SM. Hyperspectral narrow wavebands selection for optimizing water quality monitoring on the great Miami River, Ohio. Journal of Spatial Hydrology. 2001;**1**(1):1-22

[46] Helder D, Markham B, Morfitt R, Storey J, Barsi J, Gascon F, et al. Observations and recommendations for the calibration of Landsat 8 OLI and sentinel 2 MSI for improved data interoperability. Remote Sensing. 2018;**10**(9):1340

[47] Ortega S, Fabelo H, Iakovidis DK, Koulaouzidis A, Callico GM. Use of Hyperspectral/multispectral imaging in gastroenterology. Shedding some–different–light into the dark. Journal of Clinical Medicine. 2019;**8**(1):36

[48] Jensen D, Simard M, Cavanaugh K, Sheng Y, Fichot CG, Pavelsky T, et al. Improving the transferability of suspended solid estimation in wetland and deltaic waters with an empirical hyperspectral approach. Remote Sensing. 2019;**11**(13):1629

[49] Xi H, Hieronymi M, Röttgers R, Krasemann H, Qiu Z. Hyperspectral differentiation of phytoplankton taxonomic groups: A comparison between using remote sensing reflectance and absorption spectra. Remote Sensing. 2015;**7**(11):14781-14805

[50] Nelson JR, Guarda S. Particulate and dissolved spectral absorption on the continental shelf of the southeastern United States. Journal of Geophysical Research. 1995;**100**:8715-8732

[51] Vodacek A, Blough NV, DeGranpre MD, Peltzer ET, Nelson RK. Seasonal variations of CDOM and DOC in the middle Atlantic bight: Terrestrial inputs and photooxidation. Limnology and Oceanography. 1997;**42**:674-686

[52] Carder KL, Hawes SK, Baker KA, Smith RC, Steward RG, Mitchell BG. Reflectance model for quantifying chlorophyll-a in the presence of productivity degradation products. Journal of Geophysical Research. 1991;**96**:20599-20611

[53] Fell F., Fischer J, Schaale M, Schroeder T. Retrieval of chlorophyll concentration from MERIS measurements in the spectral range of sun-induced chlorophyll fluorescence. 2003. DOI: 10.1117/12.467267

[54] Kirkpatrick GJ, Orrico C, Moline MA, Matthew O, Schofield OM. Continuous hyperspectral absorption measurements of colored dissolved organic material in aquatic systems. Journal of Applied Optics. 2003;**42**(33):6564

[55] Rochelle-Newall EJ, Fisher TR, Fan C, Glibert PM. Dynamics of chromophoric dissolved organic matter and dissolved organic carbon in experimental mesocosm. International Journal of Remote Sensing. 1999;**20**(3):627-641

[56] Rochelle-Newall EJ, Fisher TR. Chromophoric dissolved organic matter and dissolved organic carbon in Chesapeake Bay. Marine Chemistry. 2002;**77**:23-41

[57] Brando V, Dekker A. Satellite hyperspectral remote sensing estimatingestuarine and coastal water quality. IEEE Transactions on Geoscience and Remote Sensing. 2003;**41**(6):1378-1381

[58] Kneubühler M, Gemperli C, Schläpfer D, Zah R, Itten K. Determination of water quality parameters in Indian ponds using remote sensing methods. In: Zagajewski B, Sobczak M, Wrzesień M, editors. Proceedings of 4th EARSeL Workshop on Imaging Spectroscopy. New Quality in Environmental Studies. Zurich Open Repository and Archive (ZORA), University of Zurich; 2005. DOI: 10.5167/uzh-97035

[59] Dingtian Y, Delu P, Xiaoyu Z, Xiaofeng Z, Xianqiang H, Shujing L. Retrieval of chlorophyll a and

suspended solid concentrations by hyperspectral remote sensing in Taihu Lake, China. Chinese Journal of Oceanology and Limnology. 2006;**24**(4):428-434

[60] Giardino C, Brando EV, Dekker GA, Strömbeck N, Candiani G. Assessment of water quality in Lake Garda (Italy) using Hyperion. Remote Sensing of Environment. 2007;**109**:183-195

[61] Giardino C, Bresciani M, Pilkaityte R, Bartoli M, Razinkovas A. In situ measurements and satellite remote sensing of case 2 waters: First results from the Curonian lagoon. Oceanologia. 2010;**52**(2):197-210

[62] Santini F, Alberotanza L, Braga F, Cavalli RM, Pignatti S. Hyperspectral applications for the improvement of water quality assessment and monitoring in coastal areas. In: Proc. 'Hyperspectral 2010 Workshop', Frascati, Italy; 17-19 March 2010 (ESA SP-683, May 2010)

[63] Chen Z, Hanson JD, Curran PJ. The form of the relationship between suspended sediment concentration and spectral reflectance: Its implication for the use of Daedalus 1268 data. International Journal of Remote Sensing. 1991;**12**:215-222

[64] Novo EMM et al. The effect of viewing geometry and wavelength on the relationship between reflectance and suspended sediment concentration. International Journal of Remote Sensing. 1989;**10**:1357-1372

[65] Xiao X, Hambright K, Zhang L, Biradar C, Puls R. Integrating in-situ field measurements and satellite remote sensing to monitor harmful algae blooms in Oklahoma lakes. 2012. Available from: http://Oklahomawatersurvey.org/?p=531

[66] Hommersom A, Kratzer S, Laanen M, Ansko I, Ligi M, Bresciani M, et al. Intercomparison in the field between the new WISP-3 and other radiometers (TriOS Ramses, ASD FieldSpec, and TACCS). Journal of Applied Remote Sensing. 2012;**6**:2012

[67] Maltese A, Capodici F, Ciraolo G, Loggia GL. Coastal zone water quality: Calibration of a water-turbidity equation for MODIS data. European Journal of Remote Sensing. 2013;**46**(1):333-347. DOI: 10.5721/EuJRS20134619

[68] Kallio K. Remote sensing as a tool for monitoring lake water quality. In: Heinonen P, Ziglio G, van der Beken A, editors. Hydrological and Limnological Aspects of Lake Monitoring. Chichester, England: John Wiley & Sons, Ltd; 2000. pp. 237-245

[69] Kong J-L, Sun X-M, Wong DW, Chen Y, Yang J, Yan Y, et al. A semi-analytical model for remote sensing retrieval of suspended sediment concentration in the Gulf of Bohai, China. Remote Sensing. 2015;7(5):5373-5397

[70] Knaeps E, Raymaekers D, Sterckx S, Odermatt D. An intercomparison of analytical inversion approaches to retrieve water quality for two distinct inland waters. In: Lacoste-Francis H, et al. editors. Proceedings of the 'Hyperspectral Workshop 2010' ESRIN, Frascati, Italy (ESA SP-683, May 2010); 2010. p. 7

[71] Lee Z-P, Arnone R, Hu C, Werdell PJ, Lubac B. Uncertainties of optical parameters and their propagations in an analytical ocean color inversion algorithm. Applied Optics. 2010;**49**:369-381

[72] Ritchie JC, Zimba PV, Everitt JH. Remote sensing techniques to assess water quality. Photogrammetric Engineering and Remote Sensing. 2003;**69**(6):695-704

[73] Reif M. Remote sensing for inland water quality monitoring: A U.S. Army Corps of Engineers Perspective Engineer Research and Development Center/ Environmental Laboratory Technical Report (ERDC/EL TR)-11-13; 2011. 44pp

[74] Dall'Olmo G, Gitelson AA. Effect of bio-optical parameter variability on the remote estimation of chlorophyll-a concentration in turbid productive water: Experimental results. Applied Optics. 2005;**44**(3):412-422

[75] Dall'Olmo G,Gitelson AA, Rundquist DC. Towards a unified approach for remote estimation of chlorophyll-a in both terrestrial vegetation and turbid productive waters. Geophysical Research Letters. 2003;**30**:1038. DOI: 10.1029/2003GL018065

[76] Gitelson AA, Gritz U, Merzlyak MN. Relationships between leaf chlorophyll content and spectral reflectance and algorithms for non-destructive chlorophyll assessment in higher plant leaves. Journal of Plant Physiology. 2003;**160**:271-282

[77] Lathrop RG. Landsat thematic mapper monitoring of turbid inland water quality. Photogrammetric Engineering and Remote Sensing. 1992;**58**:465-470

[78] Lee Z, Carder KL, Mobley CD, Steward RG, Patch JS. Hyperspectral remote sensing for shallow waters: 2. Deriving bottom depths and water properties by optimization. Applied Optics. 1999;**14**:417-427

[79] Mobley CD, Sundman LK, Davis CO, Bowles JH, Downes TV, Leathers RA, et al. Interpretation of hyperspectral remote-sensing imagery by spectrum matching and look-up tables. Applied Optics. 2005;**44**:3576-3592

[80] Cannizzaro JP, Carder KL. Estimating chlorophyll-a concentrations from remote sensing reflectance data in optically shallow waters. Remote Sensing of Environment. 2006;**101**(1):13-24

[81] Lee ZP, Carder KL, Arnone R. Deriving inherent optical properties from water color: A multi-band quasi-analytical algorithm for optically deep waters. Applied Optics. 2001;**41**:5755-5772

[82] Zhu WN, Yu Q, Tian YQ, Chen RF, Gardner GB. Estimation of chromophoric dissolved organic matter in the Mississippi and Atchafalaya river plume regions using above-surface hyperspectral remote sensing. Journal of Geophysical Research-Oceans. 2011;**116**:C02011

[83] Morillo S. Atmospheric Correction User's Guide; Laboratory for Applied Remote Sensing and Image Processing. Puerto Rico: University of Puerto Rico at Mayagüez; 2005. pp. 21-30

[84] Lee ZP, Hu C. Global distribution of Case-1 waters: An analysis from SeaWiFS measurements. Remote Sensing of Environment. 2007;**101**(2006):270-276

[85] Cao L, Li P, Zhang L, Chen T. Remote Sensing Image-Based Analysis of the Relationship Between Urban Heat Island and Vegetation Fraction. 2008

[86] Pahlevan N, Schott JR. Leveraging EO-1 to evaluate capability of new generation of Landsat sensors for coastal/inland water studies. IEEE Journal of Selected Topics in Applied Earth Observations and Remote Sensing. 2013;**6**(2):360-374

[87] Karpouzli E, Malthus T. The empirical line method for the atmospheric correction of IKONOS imagery. International Journal of Remote Sensing. 2003;**24**(5):1143-1150

[88] Gao BC, Davis CO, Goetz AFH. A review of atmospheric correction

techniques for hyperspectral remote sensing of land surfaces and ocean colour. In: Proceedings of the IEEE International Conference on Geoscience and Remote Sensing Symposium, IGARSS 2006; USA: Denver, CO; 31 July-4 August 2006. pp. 1979-1981

[89] Pahlevan N, Schott JR. Leveraging EO-1 to evaluate capability of new generation of Landsat sensors for coastal/inland water studies. IEEE Journal of Selected Topics in Applied Earth Observations and Remote Sensing. 2013;**6**(2):360-374

[90] Kutser T, Vahtmäe E, Praks J. A sun glint correction method for hyperspectral imagery containing areas with non-negligible water leaving NIR signal. Remote Sensing of Environment. 2009;**113**(10):2267-2274

[91] Lillesand TM, Johnson WL, Deuell RL, Lindstrom OM, Meisner DE. Use of Landsat data to predict the trophic state of Minnesota Lakes. Photogrammetric Engineering and Remote Sensing. 1983;**49**(2):219-229

[92] Kay S, Hedley JD, Lavender S. Sun glint correction of high and low spatial resolution images of aquatic scenes: A review of methods for visible and near-infrared wavelengths. Remote Sensing. 2009;**1**(4):697-730

[93] Hedley JD, Harborne AR, Mumby PJ. Technical note: Simple and robust removal of sun glint for mapping shallow-water benthos. International Journal of Remote Sensing. 2005;**26**(10):2107-2112

[94] Hochberg EJ, Andrefouet S, Tyler MR. Sea surface correction of high spatial resolution Ikonos images to improve bottom mapping in near-shore environments. IEEE Transactions on Geoscience and Remote Sensing. 2003;**41**(7):1724-1729

[95] Hedley JD, Harborne AR, Mumby PJ. Technical note: Simple and robust removal of sun glint for mapping shallow-water benthos. International Journal of Remote Sensing. 2005;**26**(10):2107-2112

[96] Wang Q, Li S, Jia P, Qi C, Ding F. A review of surface water quality models. The Scientific World Journal. 2013;**2013**:7

[97] Wang X, Gu X, Wu Z, Wang C. Simulation of flood inundation of Guiyang city using remote sensing, GIS, and hydrologic model. The International Archives of the Photogrammetry, Remote Sensing and Spatial Information Sciences. 2008;**XXXVII**(Part B8)